High-impact Threats to Critical Infrastructure

Conference Proceedings:

InfraGard National EMP SIG Sessions

"High-Impact Threats to Critical Infrastructure"

Emerging Policy and Technology

Conference Proceedings of the InfraGard National EMP SIG Sessions at the

Dupont Summit 2012

Background and written material for the sessions video-recorded at
http://www.ipsonet.org/conferences/the-dupont-summit/dupont-summit-2012

Edited and Introduced by Charles L. Manto

Friday, December 7, 2012
Carnegie Institution for Science
1530 P Street NW
Washington, DC

<u>High-impact Threats to Critical Infrastructure</u>

Conference Proceedings
InfraGard National EMP SIG Sessions

Dupont Summit 2012
December 7, at the Carnegie Institution for Science
1530 P Street Northwest, Washington, DC 20005
Hosted and Published by the Policy Studies Organization

Edited by Charles L. Manto, Chairman of the National EMP SIG of InfraGard®

July 2013

Table of Contents

Preface

This conference proceedings of the InfraGard EMP SIG (Electromagnetic Pulse Special Interest Group) sessions at the Dupont Summit 2012 provides written presentations and background material for the video recordings available
at http://www.ipsonet.org/conferences/the-dupont-summit/dupont-summit-2012

The sessions covered high-impact threats to critical infrastructure with a special emphasis on geomagnetic disturbance (GMD), the topic of the sessions provided by the EMP SIG at the Dupont Summit 2011 and the contingency planning workshops and exercises with the National Defense University and the Maryland Emergency Management Agency in October 2011. During the time of this conference, the Federal Energy Regulatory Commission (FERC) published a Notice of Proposed Rule-making (NOPR) on GMD. In addition to a formal presentation at the conference, a number of presenters made formal comments to the FERC, some of which have been included in these proceedings along with the NOPR so that readers of the conference material could have a background to some of the issues. On May 16, 2013, FERC issued their rule 779 on the topic which can be viewed on their website http://ferc.gov/whats-new/comm-meet/2013/051613/E-5.pdf. The appendix of this conference proceedings posted on the PSO website will provide the official ruling, other comments and links to related materials.

The EMP SIG wishes to thank the Policy Studies Organization (PSO) for its generous support of the conference and publication of these proceedings and to Mr. William Kaewert, CEO of Stored Energy Systems, LLC who provided support for transcription services of Mulberry Studios. Of course, the EMP SIG, as a nationwide special interest group of InfraGard, appreciates the strong support of the national board and staff, local chapters across the country, its members across the 50 states and three territories, and the Federal Bureau of Investigation who provides significant support to its InfraGard program.

These high-impact threats are difficult to discuss because of their grave nature and the resulting social and political concerns that they create. At the same time, the mitigating measures that many will be compelled to take could have very positive benefits for the sustainability of local communities nationwide and in any technology-enhanced society. Readers are encouraged to not only study the issue, but also join their community members to make our infrastructure as well as our relationships more sustainable as we address our mutual vulnerabilities.

As the chairman of the InfraGard EMP SIG and its conference session organizer, I welcome you to contact me for more information about InfraGard's EMP SIG and ways to participate in future activities. For information on InfraGard and how to join, see www.infragard.org..

Charles Leo Manto (cmanto@stop-EMP.com)
EMP SIG Chairman, InfraGard National
cmanto@stop-EMP.com

Introduction to the EMP SIG Sessions: Emerging Concern on High-impact Threats and Lagging Policy

The Policy Issues: High-impact low-frequency threats that could result in a month or longer collapse of critical infrastructure nationwide has been the subject of a growing number of technical and policy reports ranging from the U.S. Congressional EMP Commission, the National Academy of Sciences, Federal Energy Regulatory Commission (FERC), and NERC since 2004. This concern also led to the establishment in July of 2011 of a nationwide special interest group (SIG) in the Federal Bureau of Investigation (FBI)-sponsored program and organization known as InfraGard. This SIG that focuses on any threat that can impact critical infrastructure for more than a month was named the electromagnetic pulse (EMP) SIG after one of the four or more near-term threats that could create such a disaster (these threats include manmade or natural EMP, cyber attack, coordinated physical attack, and pandemics).

The Dupont Summit: However, nationwide contingency planning by a broad range of military and civilian government agencies and their private sector counterparts have only begun since October 2011 when the National Defense University, InfraGard's EMP SIG, and Maryland's Emergency Management Agency co-hosted a series of workshops and exercises covering these scenarios focusing on geomagnetic disturbances. In the following December, eight presenters provided an overview of the results and ramifications of those meetings at the Dupont Summit 2011 hosted by the Policy Studies Organization.

One year later, the Dupont Summit 2012 hosted sessions by InfraGard National's EMP SIG that updated activities from the prior year in a broader context of any threat that could create nationwide disruption of critical infrastructure for more than a month. This was a larger set of very densely packed half-hour sessions in a day-long conference at the Carnegie Institution for Science auditorium for over 300 registered in-person attendees and approximately 200 live webcast viewers. The presenters, an unusually broad cross-section of experts from government and the private sector by even Washington standards, provided an update of the emerging policy and technology that they are directly involved in that are engaging the threats and providing solutions.

New FBI policy and support tools were announced at the conference along with a new cyber-security awareness program by DHS. Inventors and CEOs unveiled a new technology useful in protecting the centralized grids, enhancing micro-grids, and protecting networks.

FERC and NRC: At this conference, attorney Christy Walsh of the FERC invited participants to submit comments to FERC's official Notice of Proposed Rulemaking (NOPR) on the effect of geomagnetic disturbance (GMD) on electric power grids. The comment period was open until December 24, 2012. In the same timeframe, the Nuclear Regulatory Commission (NRC) also provided their official response to a petition for rule making on the vulnerability of nuclear power plant spent fuel facilities to GMD. That petition was made by one of the conference presenters, Mr. Thomas Popik. The NRC response can be found in the Federal Register at their Vol. 77, No. 243/Tuesday, December 18, 2012/Proposed Rules pp. 74788–74797 which indicated that they will begin a phased rule-making process on the topic. These comments can be found by looking up docket number RM12-22 at ferc.gov.

Video Recordings of the Dupont Summit InfraGard EMP SIG Sessions: The Policy Studies Organization made it possible for the recording of the Dupont Summit InfraGard EMP SIG live presentations that can be viewed at their website:
http://www.ipsonet.org/conferences/the-dupont-summit/dupont-summit-2012/infragard-emp-sig

Slide presentations are also available at that link. Videos, slide presentations, and background material are also available on the secure website of the InfraGard EMP SIG. Applications for InfraGard membership can be obtained at www.InfraGard.org. Once approved for InfraGard membership, there is an EMP SIG membership application on the secure InfraGard website.

Conference Proceedings: These conference proceedings include a summary of each of the presentations and background material so that those who can view the recorded sessions may be able to place the presentations in the context of the technology policy being covered. For those who cannot view the recordings, the summaries will help fill in some of the gaps not covered by the written essays of the presenters or the transcripts of those who did not provide written essays or slides. Additional materials in the form of an on-line appendix will be available at: http://www.ipsonet.org/conferences/the-dupont-summit/dupont-summit-2012.

A summary of each presentation is given in the order of its appearance in the conference. Presenters and audience members took part in introductions and questions for presenters. A list of the presenters' biographies, a partial list of registered attendees, and pertinent FERC documents are also included.

For more information, feel free to contact the EMP SIG conference organizer and chair of the InfraGard National EMP SIG, Charles Manto, at cmanto@stop-EMP.com.

Summary of Presentations in Order of their Appearance

Dr. Pry and Congressman Roscoe Bartlett: Dr. Peter Pry, formerly a CIA analyst and staff to Congressman Bartlett and the EMP Commission he helped launch, reviewed the history of those trying to address this issue beginning with the formation of the EMP Commission to current day attempts at legislation to address the issues. He introduced Congressman Roscoe Bartlett who noted that just as people think some things are "just too good to be true", that similarly "other things are just too bad to be true," claiming that EMP and these high-impact threats are examples of the latter. Congressman Bartlett's long-term history of wrestling with high-impact threats to infrastructure such as the electric power grids parallels his long-standing congressional leadership in energy security and the role of renewable energy. Congressman Bartlett reviewed the history of his involvement of both high-impact threats to critical infrastructure with special emphasis on manmade EMP and space weather along with his related interest in energy security. He answered audience questions and led questioning of other presenters throughout the morning session.

Mr. Richard McFeely, FBI Executive Assistant Director, Criminal, Cyber, Response and Services Branch: Mr. McFeely used his address as an opportunity to not only highlight the growing concern over cyber-security, but also announced a new program that the FBI is launching to work with the private sector to protect their networks using InfraGard as a major program vehicle. The announcement included the proposed use of automated tools to assist the private sector in learning about and reporting detrimental cyber incidents.

Dr. Chris Beck of the Energy Infrastructure Security Council, provided an update of the three international conferences his organization led in the arena of extreme space weather and EMP. In each of the cases, a growing sense of concern and call for action has been noted with the Europeans appearing to be ahead of U.S. policymakers on mitigation action.

Mr. Thomas Popik, Chairman of the Foundation for Resilient Societies, outlined his interest and work as a private citizen to learn about electric grid unreliability and the need to protect it from even greater threats such as solar storms. His activity included a formal petition to the Nuclear Regulatory Commission regarding the safety of nuclear power plants in the event of a long-term grid outage. Formal announcement of the disposition of his petition is expected by mid-December 2012. He encouraged listeners to get further involved including providing comment to the current FERC call for comment on their proposed ruling that industry create standards to protect their infrastructure from geomagnetic disturbances.

Mr. "Yuki" Karakawa, Chairman of the Japan Resilience Initiative Task Force, provided an in-depth review of the nuclear disaster around the recent Fukushima Tsunami and quake. According to Mr. Karakawa, the similar resistance of government officials to fully engage this high-impact issue resulted in unnecessary deaths and continued exposure to toxic levels of radiation from the yet uncapped radioactive plants in Fukushima. He reviewed the need for local communities to be more resilient and for enhanced response capabilities. He emphasized the need for a constitutional change that would make it possible for Japan to issue an emergency declaration. All of this was not lost on the audience who assembled to hear him on this "Pearl Harbor Day."

Mr. John Kappenman, a leader or participant in many of the FERC, NERC, EPRI, Academy of Sciences and NATO studies on the issue provided a quick update of the technical issues that the industry and FERC are considering as part of the proposed ruling. He gave an orientation as to why past extreme space weather predictions often failed to distinguish the most significant impacts on power grids and how the industry has been spotty in collecting and sharing data on impacts on transformers and generators. He continues to call for fuller and transparent data collection and analysis by power utilities while taking currently available measures to protect long-lead time transformers and generators from damage. He mentioned some of the technologies that the inventors of these technologies would cover later in the afternoon. Mr. Kappenman is noted to be a consistent challenger to the industry's lack of standards and concern in this area and has borne the brunt of much of the resulting industry push-back.

Ms. Mary Lasky, Chair of the Howard County Community Emergency Response Network, spoke on local leadership's role in enhancing local sustainability in light of these high-impact threats to critical infrastructure. She heads the office of business continuity at Johns Hopkins Applied Physics Laboratory (JH APL) in Laurel, MD that hosted one of the October 2011 workshops led by the InfraGard and the Maryland Emergency Management Agency that covered the medium impact scenario for extreme space weather. She discussed the challenge of engaging extreme disaster recovery scenarios and the steps she continues to take on furthering planning steps. Her main point was that local communities need to be more self-reliant for themselves and neighbors since outside help may not be available for many months after a disaster which impacts the nation as a whole.

Mr. Chuck Manto introduced the next three speakers by noting the difficulty that federal agencies have in raising awareness of serious threats that could undermine public confidence. He used the example of the FEMA's National Preparedness Report's worst case-planning scenario published in March 2012 where local communities were asked to prepare to treat 265,000 medical casualties. Local areas were defined as areas of roughly 7 million people. Yet, each of these areas has fewer than 20,000 hospital beds leaving readers to wonder how one handles the other 245,000 casualties. Yet, no comprehensive discussion or planning for events of this magnitude has occurred; not by federal, state, or local governments, let alone the private sector. However, the recent $7B RFP by the DoD for development of local power systems for military bases, articulate and public discussions on EMP and space weather led by the Assistant Secretary of Defense, and the leadership of Dr. Andres of the National Defense University Chair of Energy and Environmental Security Programs, all appear in stark contrast to the typical hesitancy to discuss these clear and present dangers to American society and sovereignty.

Mr. Jeff Weiss, Co-chairman of Distributed Sun, is an executive who leads finance and project management of solar micro-grid projects. He also has been active in the production of the Business Executives for National Security (BENS) report on micro-grids for military bases. After mentioning the key findings of the BENS report, he introduced the next speaker from the DoD who addressed both the needs of the DoD and the reliance the DoD has on related civilian critical infrastructure.

Assistant Secretary of Defense Dr. Paul Stockton is responsible for the DoD's support of civil government and related homeland defense support. He walked through the concern that the DoD has these issues not only for military bases and the ability of the DoD to conduct its mission but also, for the country as a whole that is equally dependent on civilian critical infrastructure for its operations. He emphasized the need for military bases and local communities to become more sustainable at the local level and shared experiences of failing to do enough in these areas. He profusely thanked InfraGard for keeping these conversations active between the private sector, federal, state, and local authorities.

Dr. Richard Andres, Energy Security Program Chair of NDU, was the lead host of the first comprehensive contingency planning on a nationwide collapse of infrastructure in October 2011 with InfraGard, the Maryland Emergency Management Agency, and the U.S. House Congressional EMP Caucus. The consensus from those in these workshops and exercise was that in the event of a long-term nationwide collapse, outside help would not be available in a timely manner. Therefore, local communities must be more capable of providing essential services such as power, communications, food, water, and sanitation locally to the greatest extent possible. This creates an urgent and strong perceived need for local power generation and storage. He shared his experience in leading these conversations and noted why it was complicated. He also answered questions about how next steps might be encouraged without provoking resistance to the conversation by sounding either "too alarmist" or extreme on the issues.

Dr. George Baker, the former head of the DoD Defense Nuclear Agency's EMP Program, discussed seven persistent misconceptions about EMP as he expressed concern for the two primary critical infrastructures vulnerable to EMP and solar storms—electric power grids and telecommunications networks. Among the misconceptions were the tendency to either dismiss the threat altogether or exaggerate it. The fact-based middle ground position that shows substantial vulnerability of critical infrastructure can be demonstrated through science and engineering and is still in urgent need of mitigation. He emphasized the point made by others with technical background that engineering practices can be used to minimize both the effects of a solar storm and manmade EMP with little additional costs compared to trying to solve just one of those vulnerabilities alone. The slower geomagnetic storm pulses that create ground induced currents that can travel up through ground wires to short out transformers and other equipment are comparable to the high-altitude nuclear burst creation of E-3 pulses, whereas nuclear burst EMP E-2 pulses are similar to lightning and E-1 are nanosecond broadband pulses that can couple to equipment directly through air or through conductors such as power or communications wires. Competent engineering design can often make use of protection methods that can also protect against the E-1 pulses for relatively little additional cost.

Mr. Thomas Goldberg, Principal of Lineage Technologies, commented on the difficulty Congress had in creating cyber-security legislation and the emerging presidential executive orders that are likely to be used as stop-gap measures. He also demonstrated various cyber attacks on infrastructure, such as an alleged Georgian attack on a Russian power generator that caused it to explode and kill over 20 people working nearby. The point that impacted the audience the most was his description of how code is etched into chips at the microscopic level that allows those who produce the chips to sell access to the computers using them in such a way that passwords and network security protocols are ineffective at protecting against intrusion. The only choice presented was to either live with little to no network security or bring back chip manufacturing back to the United States (a later presenter, Curt Massey, proposed a technical solution for blocking or reducing the effectiveness of etched back doors in chips).

Ms. Christy Walsh, attorney from FERC, outlined the historic new call for a standard to protect electric power industry equipment from the effects of geomagnetic storms. She encouraged those watching to take what they learned from the conference in addition to what they experience in the private sector and local government as an opportunity to provide formal comments to FERC. FERC strongly appreciates comments from those outside of the utility industry since electric power users are usually under-represented at such calls for comment.

Dr. Fred Faxvog, a Senior Program Director from Emprimus and 40-year senior technical executive formerly of Honeywell, led a presentation from Emprimus about their new technology that can be used as protective devices for groups of transformers at substations. The cost benefits for day-to-day operations make it possible for a utility to totally recover costs in less than two years. These systems just passed independent technical assessments and are beginning to be placed into further testing and service in grids in the Midwest in coordination with the transformer manufacturer ABB. This emerging technology unveiled at the conference will have a significant impact on policy that is emerging through the FERC initiative since solutions are not only possible but, cost effective.

The Honorable James R. Woolsey, former Director of the CIA, provided extremely candid comments about the severity of the vulnerability of U.S. critical infrastructure, and the inability of most political and bureaucratic leaders to imagine how adversaries can relatively easily cause a nationwide long-term power outage. His presentation's title, "Energy and National Security: Protecting Our Ability to Use Electricity", bridged both the need to protect large critical infrastructure systems and the need to make and store electricity locally through the use of micro-grids. He provided a further foundation for inventors of local power generation technology and their security.

Cyber Education Review Panel, **Director Margaret Maxson**, DHS Office for National Cyber Security Education Strategy and **Dr. Clay Wilson**, Director for Cyber-security

Policy Studies, University of Maryland University College discussed recent advances in cyber education. Director Maxson presented her office's work in promoting cyber security awareness and education with special emphasis on the National Initiative for CyberSecurity Education (NICE) in connection with NIST that included the National Cybersecurity Workforce Framework at http://csrc.nist.gov/nice/framework/. She announced a new website that is being planned to go live the following week. Her work is intended to be a society wide initiative with industry as well as K-20 education. One hundred and fifty universities have achieved "center of excellence" certification from DHS and more are expected. Dr. Clay Wilson, reviewed the fast growing graduate program he leads at UMUC and discussed the role programs such as his can integrate into active participation in problem solving cyber and infrastructure security experienced by industry and government.

The Inventors Challenge Panel provided four very brief sets of technology that could be used to provide EMP, space weather, and cyber hardened local systems of power and communications.

Mr. Curt Massey, CEO of STT (STtealth Shield) and **Ms. Wendy Richards**, Chief Business Development Officer, Sky Catcher Solutions (Cy Key), unveiled a new combination of cyber security protection equipment. The Cy Key provides a portable systems device with its own secured private partition capable of hosting multiple operating systems that allows a user to turn any computer into a secure node that even the hosting computer can not intrude. It uses two layers of encryption and blends into the STtealth Shield technology that encrypts individual nodes on a sub-net or the Internet so that unauthorized users can "not observe or ascertain activity." They also are not responsive to port probes. These can become breakthrough elements to a smart grid system and secure communications within an EMP protected network.

Mr. John W. Spears, President, Sustainable Design Group, Inc. presented work he led as the architect for the 6400 sq.ft. Sustainable Energy Research Facility (SERF) next to Frostburg State University (FSU). SERF operates totally off-grid without any petroleum-based fuels or grid supplied electricity. SERF uses a combination of passive and active solar, geothermal, biomass, and wind turbine energy generation and energy storage in the form of heat, hot water, electric batteries, and hydrogen. (Some of the research conducted at SERF has been in conjunction with Instant Access Networks (IAN), LLC creating EMP-protected micro-grids and related shielding technology for telecommunications networks and data centers). Mr. Spears also provided examples of off-grid town homes in Frederick, MD whose total cost of ownership is less than traditional on-grid homes.

Mr. Thomas Shaw, of Homogeneous Combustion Radical Ignition Technology International, LLC (HCRITI) and **Mr. Steve Wood** (TERPS) covered two categories of emerging technology, the first being a new type of combustion technology and the second being an advance in waste-to-energy technology. Quoting from his paper discussing patents issued to Dr. Blank, "Radical ignition engines utilize chemical processes to control and enhance fuel lean combustion by storing chemicals, called radical ignition species, between the engine combustion cycles. As a result, the fuel is burned more completely and at lower peak temperatures than in conventional engines, making possible 75% or greater reduction in NOx emissions, as well as improvements in fuel economy." This also makes it possible for flex-fuel engines to "enable highly stable and low-pollution combustion in the same multi-flex fuel engine over a very wide range of such fuels, with very little cycle to cycle variation." This technology was tested at Argonne National Laboratories with support from HCRITI and IAN, LLC through a grant won through FSU and the Maryland Industrial Partnerships Program (MIPS).

The second technology area reflects a patent pending process that significantly improves on anaerobic digestion technology from Instant Access Networks, LLC (IAN), Total Energy Renewable Power Systems, LLC (TERPS), and Homogeneous Combustion Radical Ignition Technology International, LLC (HCRITI). In anaerobic digesters, bacteria break the large organic molecules in the waste down into methane. The resulting biogas can then be burned in engines to power generators for electric power supply. A series of co-inventions by IAN, TERPS, and HCRITI improve biogas yield by 30% or more compared to conventional digestion processes and reduces waste carbon. This technology thus has the potential to reduce wastewater treatment grid-dependence and/or to act as a power source for any number of desired applications, while at times being able to also sell energy back to the power grid. Anaerobic digesters are combined with a variety of other technologies, such as an emerging breakthrough engine technology, smart micro-grid controllers, and wind plus solar energy augmentations to provide reliable and cost-effective electricity.

Dr. Paul J. Rich, President, Policy Studies Organization (PSO), provided the audience with an overview of the work of the PSO, its Dupont Summit and the historic ties to the very auditorium of the Carnegie Institution for Science where these sessions took place. The Summit brings leaders from government, business, and the citizenry to discuss technology and science impacting critical infrastructure, national security, research, education, and innovation along with their social implications. Dr. Rich noted how the topics and presenters of the InfraGard sessions represented pragmatic policy and technology issues fit well into the model embraced by the PSO and welcomed the group to continued participation in future Dupont Summit conferences.

Dr. Bill Joyce, Chairman and CEO of Advanced Fusion Systems, and former CEO of Dow Chemical presented the most ambitious private sector EMP testing and manufacturing facility in development. He leads the management and investment of $60M into this state-of-the-art facility that will produce vacuum tube technology that will assist electric grid operators to protect not only against ground induced currents from solar storms or the E-3 pulses of HEMP, but also against the E-1 pulses of EMP whether produced by high-altitude nuclear burst covering large areas or pulsing devices that can disrupt or damage equipment at specific locations. The facility will be "capable of testing devices at line voltages up to 1.2 million VAC or VDC, under load conditions of up to 10 MW, and in a sub-100 picosecond rise time pulsed electric field environment of >250 kV/m." One of the AFS product lines is the Bi-tron™, a bi-directional electron tube family designed for AC power electronics switching and control operations rated to 1.2 MV and current ratings in hundreds of kiloAmperes. The Bi-tron is unique not only for the broader range of frequencies (of E-1 through E-3) they protect against (especially in the low frequency range), but in the way it can reset and manage multiple events in rapid succession. The implications for protection as well as advanced weapons built on a similar technology will undoubtedly compel the revision of current military specifications for at least some EMP protection technology.

Dr. Joyce answered questions from the audience and provided a sober assessment of the need for technology solutions and his own commitment to provide some of them through the technology his firm is bringing to market.

Mr. Chuck Manto, CEO of Instant Access Networks, LLC and InfraGard National EMP SIG Chairman, closed out the sessions by reiterating the call for comments from FERC and FBI EAD Mr. McFeely. He also encouraged continued InfraGard participation of individuals from the broad range of government agencies and private sector firms represented in the conference. As an example of that continued collaboration, he announced some of the names of those serving as national advisory panel members in the EMP SIG and of those that have offered to do so pending completion of their InfraGard applications. Mr. Manto closed the session at 5:45PM and encouraged networking of attendees at the close.

Biographies of Presenters

Andres, Dr. Richard   **Email:** rich.andres@gc.ndu.edu
Dr. Andres is a Senior Fellow and Energy and Environment Security and Policy Chair at INSS of the National Defense University at Ft. McNair, Washington, DC. His current work focuses on energy and environmental security and particularly defense-related energy issues. Prior to joining INSS, Dr. Andres was a professor at Air University assigned to the Pentagon where he served as Special Advisor to the Secretary of the Air Force. He has also served as a consultant to the Office of the Secretary of Defense (during both the Clinton and Bush administrations), the Joint Chiefs of Staff, the Office of Force Transformation, U.S. Strategic Command, the Nuclear Posture Review, the Council on Foreign Relations and other organizations. His publications appear in such journals as *International Security*, the *Journal of Strategic Studies, Security Studies*, and *Joint Force Quarterly*. Dr. Andres was awarded the medal for Meritorious Civilian Service, and has received numerous academic awards and fellowships. He received his PhD from the University of California, Davis.

Sample Articles: "Volatility in the European Energy Security Framework: Addressing Ukraine-Russia Gas Pricing Disputes," *INSS Strategic Forum* (2nd Quarter 2010). "Small Nuclear Reactors for Military Installations: Capabilities, Costs, and Technological Implications," *INSS Strategic Forum* (2nd Quarter 2010).

"Energy and Environmental Insecurity," *Joint Forces Quarterly*, Vol 55 (4th Quarter 2009). "The Emerging Energy Security System," book chapter, in *Global Strategic Assessment* (National Defense University Press 2009). "The Department of Defense: New Energy Infrastructure and Fuels," book chapter, in *Global Strategic Assessment* (National Defense University Press 2009).

Baker, Dr. George   **Email:** bakergh@jmu.edu
Dr. Baker is emeritus professor of applied science at James Madison University (JMU). In addition to teaching graduate and undergraduate S&T courses at JMU, he directed the start-up and served as Technical Director of the university's Institute for Infrastructure and Information Assurance (IIIA). Much of his career was spent at the Defense Nuclear Agency (DNA) and the Defense Threat Reduction Agency (DTRA) protecting strategic systems against electromagnetic pulse (EMP) and developing protection guidelines and standards.

He led DNA's EMP research and development program during 1987–1994 and recently served as principal staff for the Congressional EMP Commission. A primary research interest stems from his experience as Director, Springfield Research Facility—a national center for critical system vulnerability assessment. He applies lessons-learned from DoD experience to critical national infrastructure assurance and community resilience. He consults in the areas of critical infrastructure protection, EMP and geomagnetic disturbance (GMD) protection, nuclear and directed energy weapon effects, and risk assessment. He presently serves on the Board of Directors of the Foundation for Resilient Societies, the Board of Advisors for the Congressional Task Force on National and Homeland Security, the JMU Research and Public Service Advisory Board, and the National Defense Industrial Association (NDIA) Homeland Security Executive Board.

Bartlett, Congressman Roscoe G.   Email: roscoegbartlett@gmail.com
Elected to serve his tenth term in the United States House of Representatives, Roscoe G. Bartlett considers himself a citizen-legislator, not a politician. Prior to his election to Congress, he pursued successful careers as a professor, research scientist and inventor, small business owner, and farmer. He was first elected in 1992 to represent Maryland's Sixth District.

In the 112th Congress, Bartlett serves as Chairman of the Tactical Air and Land Forces Subcommittee of the House Armed Services Committee. Owing to his 10 years of experience as a small business owner, he also serves on the Small Business Committee. One of three scientists in the Congress, Dr. Bartlett is also a senior member of the Science, Space and Technology Committee.

Prior to his election to the Congress, Dr. Bartlett worked for more than 20 years as a scientist and engineer on research and development programs for the military and NASA. Nineteen of his 20 patents are held by the U.S. Government for his inventions of life support equipment used by military pilots, astronauts, search and rescue personnel, and firefighters.

In 2008, *Slate* magazine applauded him as "an advocate for reducing dependency on fossil fuels." The Association for the Study of Peak Oil (ASPO-USA) created the Roscoe G. Bartlett "Speak Truth to Power" Award in his honor in 2008. It had previously awarded him the M. King Hubbert Award in 2006 for his leadership in the Congress to promote efficiency and conservation and alternative renewable sources of domestic energy to enable the United States to overcome the challenges to national security and economic prosperity of global peak oil. Congressman Bartlett is the cofounder and cochairman of the Congressional Peak Oil Caucus. He is also the cochairman of the House Renewable Energy and Energy Efficiency Caucus and Defense Energy Security Caucus. He is also a member of the Oil and National Security Caucus.

Beck, Dr. Chris   Email: chris.beck@eiscouncil.org
Chris Beck is the President of the Electric Infrastructure Security (EIS) Council. Dr. Beck is a technical and policy expert in several homeland security and national defense-related areas including critical infrastructure protection, cybersecurity, science and technology development, WMD prevention and protection, and emerging threat identification and mitigation.

Dr. Beck served as the Subcommittee Staff Director for Cybersecurity, Infrastructure Protection, and Science and Technology and was the Senior Advisor for Science and Technology for the House Committee on Homeland Security (CHS), where he worked from May 2005 to May 2011. Prior to CHS, he worked in the office of Congresswoman Loretta Sanchez for three years, beginning as a Congressional Science Fellow and then as a legislative assistant.

Before government service, Dr. Beck was a postdoctoral fellow and adjunct professor at Northeastern University. He holds a Ph.D. in physics from Tufts University (2001) and a B.S. in physics from Montana State University (1994). He served in the Marine Corps Reserve for five years (1987–1992).

Faxvog, Dr. Frederick   Email: FFaxvog@emprimus.com
Dr. Frederick R. Faxvog, B.S., M.S., Ph.D. in Electrical Engineering, University of Minnesota, has over 40 years experience in R&D and Strategic Planning. He is presently a Sr. Program Director at Emprimus LLC. At Honeywell he was Director of Marketing and Product Management for the gyro, accelerometer and inertial measurement businesses. Earlier he was the Manager of Sensor Systems at Honeywell's Systems and Research Center (SRC). Fred made significant contributions to the growth of Honeywell's guidance and navigation business as well

as other sensor-based product areas. Before joining Honeywell he was Group Leader in the Physics Department at the General Motors Research Laboratories. He is the author of 32 publications in refereed journals and six patents.

Goldberg, Mr. Tom   **Email:** tgoldberg@atsllcdc.com
Mr. Goldberg is a founder of Lineage Technologies, LLC. Prior to founding Lineage, he founded ATS, LLC, a firm devoted to defense and homeland security matters. There he worked with DOD, DHS, DOS and DOT on physical and personnel protection matters, including the Standard Embassy Design, and blast and ballistic protection for U.S. forces, and helped U.S. firms obtain federal funding for product development, and sales, raising over $2 billion since 1993. Prior to this he worked at several trade associations, served in the Reagan and the first Bush Administrations as an advisor to the Office of the National Security Advisor. Mr. Goldberg began his career in Washington on the staff of Congressman Gilbert Gude of Maryland.

Joyce, Dr. William H.   **Email:** whjoyce@advfusion.com
Dr. William H. Joyce is the Chairman and CEO of Advanced Fusion Systems, and the retired former chairman of the board and chief executive officer of Nalco Company, the leading provider of integrated water treatment and process improvement services, chemicals, and equipment programs for industrial and institutional applications. He served in those roles at Nalco from 2003 to the end of 2007.

From 2001 to 2003, William was chairman and chief executive officer of Hercules Incorporated, a global manufacturer of chemical specialties. Prior to Hercules, William was chairman, president, and chief executive officer of Union Carbide Corporation from 1996 through 2001. In 1999, he presided over the sale of Union Carbide to Dow Chemical and served as vice chair of Dow Chemical until his departure. From 1995 to 1996, William was president and chief executive officer, and from 1993 to 1995 he was president of Union Carbide. Prior to that, he had been chief operating officer of Union Carbide since 1992.

An active member of scientific bodies and panels, William has received numerous awards and honors. He was awarded the National Medal of Technology from President Clinton in 1993. Other accolades include the Perkin Medal from the Society of Chemical Industry and the Outstanding Achievement Award and Lifetime Award from the Society of Plastic Engineers. In 2008, William was selected as one of the 100 most successful engineers of the century by the American Institute of Chemical Engineers.

William served as the elected chairman of the Society of Plastics Industry and on the executive board of the American Chemistry Council.

William received a B.S. in chemical engineering from Penn State University, and an M.B.A. with distinction and Ph.D. in business, both from New York University. At graduation he received the McDowell Award as Outstanding Graduate Student relating to his doctoral studies at New York University.

Kappenman, Mr. John   **Email:** JKappenma@aol.com
Mr. Kappenman has been an active researcher on geomagnetic storms, EMP, and their disruptive effects on electric power systems. He was previously employed at Minnesota Power and with Metatech Corp. He is the past chairman of the IEEE Transmission and Distribution Committee. Mr. Kappenman was a Scientific Organizer and one of the Principal Lecturers at the NATO Advanced Science Institute on Space Storms held in June 2000 and a series of other leadership

roles in related activities at NOAA's Space Weather Prediction Center and NATO while also providing formal testimony for U.S. Congressional committees. He was a principal investigator on extreme space weather for the U.S. EMP Commission, FEMA under Executive Order 13407, and the 2008 National Academies of Science report on "Severe Space Weather". He has recently provided extensive information to the U.S. House Committee on Homeland Security regarding cyber, EMP, and geomagnetic storm threats to the U.S. power grid infrastructure. He is currently a member of the Joint U.S. Department of Energy/NERC Steering Committee for developing and planning a conference on High Impact Low Frequency (HILF) Threats to the U.S. Electric Power Grid, which will be held November 9–10 in Washington, DC.

Karakawa, Mr. Nobuyui (Yuki)   **Email:** <u>ykarakawa@mac.com</u>
Mr. Karakawa has been involved in a number of activities responding to the recent natural and nuclear disaster in Fukishima. He is part of the Delegation of the American Medical Association, Center for Public Health Preparedness and Disaster Response; the Japan Resilience Initiative, and Vice President of the Corporation for Emergency Preparedness and Response. He is adjunct faculty at Tokyo University of Science and an Assistant Dean at Florida Institute of Technology. He has been appointed as an Executive Advisor for Councilor of the National Diet of Japan (Upper house Senator). He previously served in positions to PriceWaterhouseCoopers and CAP Gemini, UK. Mr. Karakawa received his bachelor's degree from the Tokyo University of Science, Electronics Department/Medical Engineering in 1991.

Lasky, Ms. Mary   **Email:** <u>mary.lasky@jhuapl.edu</u>
Ms. Lasky volunteers as the Chairman of the Howard County Community Emergency Response Network (CERN) and currently is Program Manager for Business Continuity Planning for the Johns Hopkins University Applied Physics Laboratory (JHU/APL). She leads APL business continuity planning including their pandemic influenza response effort. She also coordinates the APL Incident Management Team, which is responsible for managing any disaster that disrupts normal operations for more than 48 hours. CERN's mission is to spearhead the development of a community-wide disaster response plan to ensure maximum preparedness in the event of a terrorist attack or major natural emergency. The unique community readiness program was initiated after the events of 9/11 as a partnership between the Horizon Foundation, Howard County government, and key community agencies in Howard County. The effort supports government disaster planning through coordination of the emergency plans and resources of participating members. CERN functions include planning, a high level of inter-agency coordination, the development of tabletop exercises, disaster plan review, shelter planning, and communications enhancement. In addition to her APL roles, Lasky is the chair of the Pandemic Influenza Education and Communication Group for the Johns Hopkins Office of Critical Event Preparedness and Response (CEPAR), which is responsible for pandemic planning for all Hopkins Institutions. She is also on the adjunct faculty of the Johns Hopkins University, Whiting School of Engineering, teaching in the Technical Management graduate degree program.

Manto, Mr. Charles   **Email:** <u>cmanto@stop-EMP.com</u>
Mr. Manto is CEO of Instant Access Networks, LLC a consulting and R&D firm that produced independently tested solutions for EMP protected micro-grids. He received five patents in telecommunications and computer mass storage, has others pending (in advanced micro-grids and EMP protection), and assisted other entrepreneurs and investors with their intellectual

property strategies. Developed valuation methodology accepted by the U.S. DOD, countries, and companies participating in industrial defense conversion. Facilitated due diligence of over 200 deals, managed a venture capital service, a revolving loan fund, an economic development corporation, a computer mass storage manufacturer, and broadband CLEC. Mr. Manto has also founded and leads InfraGard National's EMP SIG. He received his B.A. and M.A. from the University of IL at Urbana/Champaign.

Massey, A. Curt   **Email: alancmassey@gmail.com**
Chief Executive Officer Founding Partner, of STT, LLC of Austin, TX. Curt Massey has devoted a 35-year career to protecting national security. A visionary and entrepreneur, Curt leads STT strategic direction, advancing company mission and organizational growth.

Maxson, Ms. Margaret (Peggy)   **Email: Margaret.Maxson@hq.dhs.gov**
Director, National Cybersecurity Education Strategy, DHS DHS/NPPD/NCSD/CEO. On April 19, 2010, Ms. Maxson was appointed to her position as Director of National Cybersecurity Education Strategy at the Department of Homeland Security. In this capacity she leads DHS efforts to build capability within the National Initiative for Cybersecurity Education (NICE) as well as co-leading the training and professional development component of the initiative. DHS requested Ms. Maxson for this position following her previous position at the Office of the Director of National Intelligence, when she led a cybersecurity education sub-group of the White House, which resulted in the accepted recommendation and subsequent implementation of the establishment of NICE. Ms. Maxson served for over 35 years at the National Security Agency in managerial positions in operations, policy, foreign relations, customer service, and technology development.

McFeely, Mr. Richard A.   **Email: richard.mcfeely@ic.fbi.gov**
Director Robert S. Mueller III has named Richard A. McFeely as the executive assistant director of the Criminal, Cyber, Response Services Branch. Mr. McFeely most recently served as the special agent in charge of the Baltimore Field Office.

Mr. McFeely entered on duty as a special agent in February 1990 in the Buffalo Division, where he worked violent crimes. He also served as the lead case agent investigating the Oklahoma City federal building bombing in 1995.

In 1997, Mr. McFeely was promoted to supervisory special agent in the Criminal Investigative Division at FBI Headquarters. In this role, he oversaw the FBI's drug investigations. A year later, he was detailed to the Executive Office of the President, Office of National Drug Control Policy, where he helped formulate national policy on the nation's counter-narcotics efforts.

In 1999, Mr. McFeely transferred to the Washington Field Office. Following the September 11, 2001 attack on the Pentagon, he assumed the role of the FBI's on-scene commander. Afterward, he supervised a WFO Counterterrorism Squad and was instrumental in setting up a joint intelligence center with the Fairfax County, Virginia police, and other local agencies to increase information sharing.

He received the Director's Award in 2005 for Outstanding Counterterrorism Investigation for supervising a multi-national investigation into an assassination plot against a foreign head of state. Later that year, he became the assistant special agent in charge responsible for all technical programs.

In 2006, Mr. McFeely returned to Headquarters as a section chief in the Criminal Investigative Division. He was responsible for undercover programs before becoming the FBI's budget officer. Later, he was promoted to deputy chief financial officer with oversight of the FBI's budget, acquisition, and accounting functions.

Mr. McFeely has a Bachelor of Science degree in criminal justice from the University of Delaware and a Juris Doctor degree from the Delaware Law School.

Popik, Mr. Thomas (Tom)   **Email:** thomasp@resilientsocieties.org

Thomas Popik is chairman of the Foundation for Resilient Societies, a nonprofit group dedicated to the protection of critical infrastructure against infrequently occurring natural and manmade disasters. He is principal author of a Petition for Rulemaking submitted to the Nuclear Regulatory Commission that would require backup power sources for spent fuel pools at nuclear power plants. Previously, as a U.S. Air Force officer, Mr. Popik investigated unattended power systems for remote military installations. Mr. Popik graduated from MIT with a B.S. in mechanical engineering and from Harvard Business School with an M.B.A.

Pry, Dr. Peter Vincent   **Email:** peterpry@verizon.net

Dr. Pry is the Executive Director of the Task Force on National Homeland Security. He has served: on the Commission on the Strategic Posture of the United States established by the U.S. Congress (2008–2009); as Director of the United States Nuclear Strategy Forum, an advisory body to Congress on policies to counter Weapons of Mass Destruction (2005–2009); on the Commission to Assess the Threat to the United States from Electromagnetic Pulse (EMP) Attack (also commonly known as the EMP Commission), established by the U.S. Congress (2001–2008); as Professional Staff on the House Armed Services Committee of the U.S. Congress, with portfolios in nuclear strategy, WMD, Russia, China, NATO, the Middle East, intelligence, and terrorism (1995–2001); as an Intelligence Officer with the Central Intelligence Agency responsible for analyzing Soviet and Russian nuclear strategy and operational plans (1985–1995), where he was formally recognized by the agency for his expertise, groundbreaking research, and his outstanding accomplishments during his 10 years of service; and as a Verification Analyst at the U.S. Arms Control and Disarmament Agency responsible for assessing Soviet compliance with nuclear and strategic forces arms control treaties (1984–1985). Dr. Pry also played a key role: running hearings in Congress that warned about how terrorists and rogue states could pose an EMP threat, establishing the Congressional EMP Commission, helping the Commission develop plans to protect the United States from EMP, and working closely with senior scientists who first discovered the nuclear EMP phenomenon. Dr. Pry holds two Ph.D.s (in International Relations and U.S. History) and a certificate in nuclear weapons design from the USAF Weapons Laboratory. He has also written numerous books on national security issues.

Rich, Dr. Paul   **Email:** pauljrich@gmail.com

President, Policy Studies Organization, Washington, D.C. Adjunct Professor, George Mason University. Visiting Fellow, Hoover Institution, Stanford University.

As President of the Policy Studies Organization, Dr. Rich is responsible for appointing the editors of 12 journals and of book series published for the PSO by Wiley-Blackwell, Berkeley Electronic Press, Rowland & Littlefield, and Global Information Company. The journals and their editorial offices include Policy and Internet (University of Oxford), Review of

Policy Research (Ryerson University, Toronto), Digest of Middle East Studies (University of Wisconsin), Risk, Hazards & Crisis in Public Policy (Pacific Disaster Center, University of Hawaii), Policy Studies Journal (University of Colorado, Denver), Politics & Policy (Universidad IberoAmericana, Mexico City), Latin American Policy (Tecnologico de Monterrey, Mexico City), Asian Politics & Policy (Asian Center, University of the Philippines), World Medical and Health Policy (George Mason University), Poverty & Public Policy (University of Missouri, Kansas City), Proceedings of the PSO (PSO Headquarters, Washington, DC).

The PSO publishes *The Yearbook of Policy Studies* (University of Oklahoma), book series on *Public Policy, The Middle East and China* (University of San Francisco), and hosts the Dupont Summit on Science & Technology Policy—an annual December conference in Washington DC at the Carnegie Institution, the Middle East Dialogue at the Washington Club each February, practitioner workshops at annual meetings of APSA, Westminster style parliamentary debates in cooperation with Tulane University, and as a sponsor with Oxford University's Internet Institute of biannual internet policy world conference.

The Policy Studies Organization includes more than 3,600 universities and institutions as well as individuals in more than 93 countries. From its headquarters in an historic house near Dupont Circle in Washington, home of the families of President James Garfield and the labor leader Samuel Gompers, the society organizes conferences, seminars, the publication of journals and books, and an increasing Internet presence. The PSO offices have a notable antiquarian collection relating to its history and the Dupont Circle area. Interns are placed with PSO by Cornell, Penn State, and the Washington Center; PSO visiting professors are provided office space.

Dr. Rich is a Kinsmen Scholar, Tonbridge School, England. Jr. Common Room Chr., Dunster House, Harvard College.AB cum laude, EDM, Harvard.Ph.D., University of Western Australia. Harvard Mountaineering Club (life). Secy. Harvard Liberal Union.

Richards, Ms. Wendy   **Email:** wendy.richards@merlincryption.com
Ms. Richards is VP of Business Development for Sky Catcher Solutions a company developing cyber security products. She has extensive experience in telecommunications services, healthcare, and security applications. She will be co-presenting with Mr. Curt Massey on breakthroughs on a "hyper-secure network."

Shaw, Mr. Thomas   **Email:** topaz0@gmail.com
Mr. Shaw, a recent physics and engineering graduate, is working with a team of leading researchers, including Dr. David A. Blank, on next generation combustion and waste-to-energy projects. His team at HCRTI is presenting a set of emerging technologies developed in part with Frostburg State University and IAN, LLC.

Soysal, Mrs. Hilkat   **Email:** hilkatoguz@me.com
Hilkat S. Soysal received a Law degree from University of Istanbul, Turkey. She practiced law in private companies and two state universities as a counselor. In 1993, she joined Istanbul University College of Engineering as a Lecturer. While teaching law courses for undergraduate engineering students, she completed her graduate study in the Marine Engineering program and received a M.Sc. degree in 1996. Since fall 2000, she has been with the Department of Physics and Engineering at Frostburg State University, where she is currently a Lecturer and director or co-director of several renewable energy projects including WISE Education Program, Hydrogen

Collection and Storage for Power Systems, and Sustainable Energy Research Facility (SERF).

Soysal, Dr. Oguz   **Email:** OSoysal@frostburg.edu

Oguz A. Soysal received the B.Sc., M.Sc., and Ph.D. degrees from Istanbul Technical University, Turkey. In 1983 he joined ABB-ESAS Power Transformer Company (Istanbul, Turkey) as an R&D engineer. From 1986 to 1993 he worked for Black-Sea Technical University, Turkey. In 1987 he visited The Ohio State University (OSU) as a Post Doctoral Scholar, and in 1991–1992 he spent a sabbatical leave at the University of Toronto. Between 1993 and 1997 he has been with Istanbul University, Turkey, and Bucknell University, Lewisburg, PA, USA. He is currently professor at Frostburg State University Physics and Engineering Department. His field of research includes power engineering, renewable energy, and electrical engineering education. He directed several renewable energy projects and currently he is co-director of the Sustainable Energy Research Facility (SERF).

Spears, Mr. John   **Email:** john@sustainabledesign.com

Mr. John Spears, president of the Sustainable Design Group in Gaithersburg, Md. is the architect of what is believed to be the nation's only totally off-grid building dedicated to researching renewable energy, the Sustainable Energy Research Facility.

The facility is powered, heated, and cooled by a variety of renewable energy sources, including a 10-kilowatt wind turbine, 20 kilowatts of solar photovoltaic, a solar thermal system, passive solar, geothermal and hydrogen fuel cells. The design also includes many energy conservation and energy storage features.

"The biggest challenge with this building is you have no backup because the objective was to use no fossil fuels at all," Mr. Spears notes. "There is no generator. There is no plug we can plug into if the batteries which are powered by solar and wind, run out. Therefore, we really, really had to be careful of everything that uses energy in the building, then do that in the most efficient way and sometimes in very innovative ways to avoid the need for electricity."

Stockton, Dr. Paul N.   **Email:** pstockton@cloudpeak.sonecon.com

Assistant Secretary of Defense for Homeland Defense and Americas' Security Affairs

Paul N. Stockton was nominated by President Barack Obama to be the Assistant Secretary of Defense for Homeland Defense and Americas' Security Affairs on April 28, 2009, and was confirmed by the Senate on May 18, 2009. In this position, he is responsible for supervising the Department of Defense's homeland defense activities (including Defense Critical Infrastructure Protection and other mission assurance efforts), defense support of civil authorities, domestic crisis management, and Western Hemisphere security matters.

Assistant Secretary Stockton received a bachelor's degree from Dartmouth College Summa Cum Laude in 1976, and a doctorate in government from Harvard in 1986. From 1986 to 1989, Assistant Secretary Stockton served as legislative assistant to Senator Daniel Patrick Moynihan, advising the senator on defense, intelligence, and counter-narcotics policy, and serving as the Senator's personal representative to the Senate Foreign Relations Committee. From 1989 to 1990, Assistant Secretary Stockton was a Postdoctoral Fellow at Stanford University's Center for International Security and Cooperation. During his graduate studies at Harvard, he served as a research associate at the International Institute for Strategic Studies in London.

Walsh, Ms. Christy   **Email: christy.walsh@ferc.gov**
Christy Walsh is a Special Counsel in the Office of General Counsel at the U.S. Federal Energy Regulatory Commission. She has been with the Commission since 2003, first as an attorney–advisor and senior attorney in the Office of General Counsel then serving as a Legal and Policy Advisor to Chairman Jon Wellinghoff. She worked extensively on issues involving the Commission's reliability authority, including cyber and physical security and issues involving the various electric and gas industries. She holds a J.D. from the University of Texas School of Law and a B.M. and B.A. from Texas Tech University.

Weiss, Mr. Jeff   **Email: jeff@distributedsun.com**
Jeff Weiss is co-founder and Managing Director of D-Sun's Solar Energy Investment Companies (SEIC's). Mr. Weiss leads capital formation activities for SEIC's, the entities that own and operate the company's solar assets. He also plays an active management and oversight role at D-SUN. Mr. Weiss has founded, managed, and led many companies as General Manager, CFO, CMO, Board Member and venture investor. Among them are Trust Strategy Group, a $10MM strategic intelligence firm, Picture Network International (sold to Kodak in 1997), CDx (Certificate of Deposit Exchange), and Vista Information Technologies (a $100MM network services firm).

Woolsey, Honorable R. James   **Email: jim@woolseypartners.com**
R. James Woolsey is Chairman of Woolsey Partners LLC and a Venture Partner with Lux Capital Management. He also chairs the Board of the Foundation for Defense of Democracies and is a co-founder of the United States Energy Security Council.

Mr. Woolsey currently chairs the Strategic Advisory Group of the Washington, DC private equity fund, Paladin Capital Group, chairs the Advisory Board of the Opportunities Development Group, and he is Of Counsel to the Washington, DC office of the Boston-based law firm, Goodwin Procter. In the above capacities he specializes in a range of alternative energy and security issues.

Mr. Woolsey previously served in the U.S. Government on five different occasions, where he held Presidential appointments in two Republican and two Democratic administrations, most recently (1993–1995) as Director of Central Intelligence. From July 2002 to March 2008 Mr. Woolsey was a Vice President and officer of Booz Allen Hamilton, and then a Venture Partner with VantagePoint Venture Partners of San Bruno, California until January 2011. He was also previously a partner at the law firm of Shea & Gardner in Washington, DC, now Goodwin Procter, where he practiced for 22 years in the fields of civil litigation, arbitration, and mediation.

During his 12 years of government service, in addition to heading the CIA and the Intelligence Community, Mr. Woolsey was: Ambassador to the Negotiation on Conventional Armed Forces in Europe (CFE), Vienna, 1989–1991; Under Secretary of the Navy, 1977–1979; and General Counsel to the U.S. Senate Committee on Armed Services, 1970–1973. He was also appointed by the President to serve on a part-time basis in Geneva, Switzerland, 1983–1986, as Delegate at Large to the U.S.–Soviet Strategic Arms Reduction Talks (START) and Nuclear and Space Arms Talks (NST). As an officer in the U.S. Army, he was an adviser on the U.S. Delegation to the Strategic Arms Limitation Talks (SALT I), Helsinki and Vienna, 1969–1970.

Mr. Woolsey serves on a range of government, corporate, and non-profit advisory boards and chairs several, including that of the Washington firm, ExecutiveAction LLC. He serves on the National Commission on Energy Policy. He is currently Co-Chairman of the Committee on the Present Danger. He is Chairman of the Advisory Boards of the Clean Fuels Foundation and the New Uses Council, and a Trustee of the Center for Strategic & Budgetary Assessments. Previously he was Chairman of the Executive Committee of the Board of Regents of The Smithsonian Institution, and a trustee of Stanford University. He has also been a member of The National Commission on Terrorism, 1999–2000; The Commission to Assess the Ballistic Missile Threat to the U.S. (Rumsfeld Commission), 1998; The President's Commission on Federal Ethics Law Reform, 1989; The President's Blue Ribbon Commission on Defense Management (Packard Commission), 1985–1986; and The President's Commission on Strategic Forces (Scowcroft Commission), 1983.

Mr. Woolsey has served in the past as a member of boards of directors of a number of publicly and privately held companies, generally in fields related to technology and security, including Martin Marietta; British Aerospace, Inc.; Fairchild Industries; and Yurie Systems, Inc. In 2009, he was the Annenberg Distinguished Visiting Fellow at the Hoover Institution at Stanford University and in 2010–2011 he was a Senior Fellow at Yale University's Jackson Institute for Global Affairs.

Mr. Woolsey was born in Tulsa, Oklahoma, and attended Tulsa public schools, graduating from Tulsa Central High School. He received his B.A. degree from Stanford University (1963, With Great Distinction, Phi Beta Kappa), an M.A. from Oxford University (Rhodes Scholar 1963–1965), and an LL.B from Yale Law School (1968, Managing Editor of the Yale Law Journal).

Mr. Woolsey is a frequent contributor of articles to major publications, and from time to time gives public speeches and media interviews on the subjects of energy, foreign affairs, defense, and intelligence. He is married to Suzanne Haley Woolsey and they have three sons, Robert, Daniel, and Benjamin.

For Immediate Release
Monday September 20, 2011
Contact: Chuck Manto
(410) 991-1469

PRESS RELEASE:
InfraGard Launches EMP Special Interest Group

Washington, DC – InfraGard announces the launch of a nationwide special interest group (SIG) that will focus on threats that could cause nationwide long-term critical infrastructure collapse. Named the EMP SIG, after electromagnetic pulse, the SIG will cover all similarly dangerous hazards such as extreme space weather, coordinated physical attack, cyber attack, or pandemics.

One of the SIG's first activities will be participation (in conjunction with the U.S. Congressional EMP Caucus and the National Defense University) in a nationwide conference, "Severe Space Weather Threats to the U.S. Electrical Grids" planned for October 6, 2011 from 8.30 AM to 3 PM at the U.S. Capitol Visitors Center Auditorium. This conference will provide initial reports on planning workshops and an exercise representing the first comprehensive efforts by federal, state, local government, and the private sector to examine the potential for prolonged nationwide power outages and their cascading effects and begin mitigation efforts. The InfraGard National EMP SIG plans to assist in organizing private sector critical infrastructure stakeholders who wish to mitigate these threats and make their local communities more resilient.

To register for the October 6 conference, see http://solarexercise.eventbrite.com. Any InfraGard member can join the EMP SIG. If you are not an InfraGard member you can join via the website at www.InfraGard.org and clicking on the "Apply for Membership" button. There is no charge for membership in InfraGard or the EMP SIG. Conference participation is also free-of-charge because of the generous support by SAIC. For additional information on the InfraGard EMP SIG feel free to email the EMP SIG manager, Chuck Manto at cmanto@stop-EMP.com or InfraGard National Members Alliance Managing Director Stephen Ewell at sewell@infragardnational.org.

About InfraGard and the InfraGard National Members Alliance
The InfraGard Program began in 1996 as a collaborative effort between private sector cyber professionals and the FBI field office in Cleveland Ohio. The FBI later expanded the program to each of the FBI's 56 field offices. In 2003 the private sector members of InfraGard formed the "InfraGard National Members Alliance" (INMA). The INMA is a nonprofit Delaware LLC with 501(c)3 status. The INMA LLC is comprised of 86 separate 501(c)3 InfraGard Member Alliances (IMAs) representing over 45,000 FBI-vetted, InfraGard Subject Matter Experts. The INMA has a dual-focus value proposition. First, InfraGard provides its members with unmatched opportunities to promote the physical and cyber security of their organizations, through access to a trusted, national network of Subject Matter Experts from the public and private sectors. Secondly, it provides government stakeholders, at the local, state, and Federal levels, with unmatched access to the expertise and experience of critical infrastructure owners and operators.

For more information about InfraGard, please visit www.infragard.net. For more information about the INMA, please visit www.infragardmembers.org.

The purpose

The purpose of the InfraGard National EMP (electromagnetic pulse) SIG (special interest group) is to address and mitigate the threat of a simultaneous nationwide collapse of infrastructure from any hazard such as manmade or natural EMP. Any threat that could cause a similar collapse of infrastructure over most or all of the United States is also of interest to the EMP SIG.

Method Focusing on Local Sustainability

The National EMP SIG will mobilize resources at the national level so that the local InfraGard chapters and local EMP SIGs can make use of them to help local communities become more sustainable in light of these threats.

Resources

Expert National Advisory and Liaison Panels:

The National EMP SIG will recruit boards or panels of leading advisors in various subject matter covering specific threats (such as EMP and extreme space weather), all impacted critical infrastructures, and critical relationships and communications between stakeholders. Given that many if not all of these resources may be voluntary in nature or minimally funded, it is intended that these resources become available on a best efforts basis to the local chapters needing guidance. Critical relationships can be supported by the EMP SIG through liaison panels. Examples may include:

1. a **civilian–military liaison panel** enabling local communities to become more resilient so that they can better support their local DoD and National Guard resources;
2. a **legislative and policy liaison panel** that can facilitate discussions between interested leaders at the national and local level with those in the private sector to identify and fill policy gaps;
3. an **education panel** that can facilitate research, development, and education/training into the development of human resources needed;
4. an **investment panel** that might identify the capital support needed by local communities to enhance its sustainability;
5. a **media, communications, and outreach panel** that can facilitate communications among EMP SIG members and between other stakeholders outside the SIG; and
6. an **InfraGard liaison panel** that would coordinate activities between the EMP and other SIGs and committees within InfraGard.

Qualifications and Expectations of National Panel Members

Qualifications: Panel members will be chosen based on their leadership within their respective fields by virtue of knowledge, experience, capabilities, or relationships.

Expectations: Panel members will agree to an in-person meeting once per year, several conference calls during the course of the year and occasional email or phone correspondence between themselves and EMP SIG leadership. However, given the volunteer nature of these roles, it is expected that their contributions while meaningful will be limited and on a best-efforts basis.

Appointment of National Panel Members: Membership to the national panels will be by appointment by the EMP SIG manager or designated Federal Bureau of Investigation (FBI) SIG liaison and can be revoked either by the SIG manager, EMP Guidance Committee, or FBI SIG liaison. The initial chairman of each panel will be appointed by the EMP SIG manager and subsequently by vote of the panel themselves at intervals of their selection.

The InfraGard National EMP SIG leadership team includes:
1. The InfraGard National EMP SIG manager who serves at the pleasure of the InfraGard National Members Alliance Board of Directors.
2. The InfraGard Guidance Committee composed of the InfraGard National Members Alliance Board Chairperson, the InfraGard National Members Alliance Managing Director, and an FBI HQ Supervisory Special Agent from the Public/Private Alliance Unit (PPAU) who functions as an official liaison to the FBI for InfraGard.

EMP SIG Membership
Any InfraGard member in good standing may join the EMP SIG at the national level by indicating interest in participating in activities, mailings, or communications designed for membership participation. EMP SIG members are encouraged to ask EMP SIG national leadership for help with their local EMP SIG activities.

Secure Communications between SIG Members
InfraGard intends to make its secure Intranet resources available to the EMP SIG as they become available. EMP SIG leadership will also assist in providing other resources to supplement the official InfraGard resources as needed on a best-efforts basis. This will include library resources and links to resources deemed to be of special value by the membership of the EMP SIG.

Other Resources
The EMP SIG will be responsible to recruit and raise resources necessary to do its tasks subject to the normal and customary procedures and governance of InfraGard National Members Alliance and Chapter organizations unless specifically authorized by InfraGard.

Governance and Activities of the EMP SIG
All activities of the EMP SIG will comply with the governance and ethics laid out by the InfraGard organization and any guidance provided by the InfraGard Board of Directors.
(Background: The concepts of this guidance document have been proposed by the founding SIG manager and approved by the InfraGard National leadership and the FBI. See the initial authorizing letter for the EMP SIG and the initial EMP Committee/SIG proposal for additional background).

ROSCOE G. BARTLETT
8TH DISTRICT, MARYLAND

2412 RAYBURN HOUSE OFFICE BUILDING
WASHINGTON, DC 20515
(202) 225-2721

**UNITED STATES
HOUSE OF REPRESENTATIVES**

April 5, 2012

Mr. Charles Manto, CEO
Instant Access Networks, LLC
230 Baltimore Ave.
Cumberland, MD 21502

Dear Mr. Manto:

I want to personally congratulate you and express appreciation for the development of the new nationwide InfraGard EMP SIG (electromagnetic pulse special interest group). Your initiation and help in organizing the historic contingency planning work on space weather planning of the National Defense University (NDU) and the Maryland Emergency Management Agency at NDU October 3-5 and the US Capitol Visitor Center in Washington, D.C. on October 6, 2011 was instrumental in its success. InfraGard also supported the follow-on academic conference of the Policy Studies Organization on December 2, 2011 in Washington, D.C.

Up until this series of workshops and exercise on October 3-6, no contingency planning had been developed for a nationwide collapse of critical infrastructure which could last more than a few weeks. InfraGard support was not only timely then, but, I know of no other organization that can foster future trusted conversations among the nation's leading providers and protectors of our critical infrastructure to work through these issues.

As the nation addresses high-impact threats that range from the hundred-year solar storm to cyber attacks, successful contingency planning and mitigation will ultimately depend on how local communities will become more sustainable. One of the key ways to foster local sustainability would be for each community to make 20% of their own power. More often than not, local power will use renewable energy with all the environmental and energy security benefits that result. This addresses a strategic convergence of energy security and critical infrastructure vulnerability that reflects my own long-term work in Congress. To be successful, it is critical to engage public and private sector leaders across all critical infrastructure sector. I trust that InfraGard will continue to take a leading role in helping our nation to competently confront these threats.

Again, thank you for your volunteer support of InfraGard and I look forward to more good things to come.

Sincerely,

ROSCOE G. BARTLETT
Member of Congress

Welcome and Introductory Remarks
Charles Manto, Chairman, InfraGard National EMP SIG

Welcome to the InfraGard National Electromagnetic Pulse Special Interest Group (EMP SIG) sessions at the Dupont Summit for 2012. I am Chuck Manto, founder and current volunteer chairman of the EMP SIG, the most active nationwide special interest group in InfraGard today.

InfraGard is a Federal Bureau of Investigation (FBI) program that began in 1996 and has grown to a nonprofit association of roughly 50,000 individuals in public and private sectors who are concerned about critical infrastructure and its protection. In order to join, members sign NDA's with each other and the FBI to make it possible to have trusted conversations about difficult topics—in particular—our mutual concerns about vulnerabilities of the infrastructure we all depend on. This combination of NDAs and background checks is the closest thing to a civilian security clearance that the nation currently has available. Lack of trust often prevents many private sector organizations from collaborating with each other and government agencies on critical issues vital to our security. InfraGard is unique in its ability to help solve that trust problem and provides the opportunity for dialog necessary for effective infrastructure security and resilience.

Formed just last year, the InfraGard National EMP SIG has members in each of the 50 states and three territories focusing on ANY threat that can disrupt critical infrastructure nationwide for more than a month. Its primary focus has been on man-made and natural EMP, cyber attacks, or coordinated physical attack. Any other threat that can cause a long-term nationwide collapse of infrastructure, such as pandemics or economic collapse, would also be of interest. The EMP SIG takes a truly all-hazards approach to the topic.

Today we are fortunate to again be hosted by the Policy Studies Organization (PSO)—an organization of universities that focuses on public policy—particularly in the areas of science, technology, and the environment. Each first Friday of December, they convene the Dupont Summit conference where academic, business, and government leaders share ideas that promote dialog on pressing policy issues. We appreciate the support of the PSO, and their sponsors, along with InfraGard and the FBI. Without their support this conference would not be possible.

Today's sessions will be webcast and recorded. Within a few months, we should also have a conference proceedings publication that will go into a number of topics that will be covered today in far more detail than we have time for in just one day of sessions.

Leadership Issues
As we begin, I would like to mention that those of us who work in these areas understand the great difficulty we face in trying to look at anything so depressing and disorienting as the prospect of months-long collapse of critical infrastructure. In the event of such a calamity, the loss to the economy and lives can be so great that there is a temptation to just not want to consider the "unthinkable." Politically, it is not considered wise to upset the public. From a management perspective, it has not been anyone's job to work on these issues. There are no resources set aside to deal with this. So, it is not unusual for anyone speaking about this topic to

his or her peers to hear, "How could this be a real problem if I have not heard about it?" This category of threats are really challenging for both our leadership as well as the average citizen.

Today you will hear from top technical and policy experts who are shaping emerging policy and technology in this area despite this difficulty. I hope you will appreciate the work and sacrifice that they have made in order to bring these issues to our attention.

In a way, these high-impact low-frequency topics are much like this year's hurricane Sandy. We can all be so overwhelmed and under-resourced in our work and even our personal lives, that we can hardly think of trying to deal with anything more than we already have on our plates. But, if the storm is coming, it doesn't matter how hard the economy is or how tired we are. We still need to do what we can to prepare for the coming storm. So, as you think about the issues the presenters will speak about today and seek to engage them and each other today and the days that follow, I encourage you to think about the coming storm and ask what you can do to be better prepared.

The first two speakers, as others on today's panels, have spent many years working on these issues. Dr. Pry will speak in a moment and introduce Congressman Bartlett who not only led congress on these issues, but, in the words of Senator Cardin, spoke to Congress about energy issues before they knew how to say "peak oil."

Dr. Pry, you have had a long and distinguished career in service at the CIA and as congressional staff on the very issues we are facing today and we welcome you as you give us a brief history and update on the EMP scenarios outlining both the good news and remaining concerns. Let's welcome Dr. Pry…

(Dr. Pry takes the stage.)

Too Little, Too Late? Speech on the Status of National EMP Preparedness Before the FBI
Peter Vincent Pry, Executive Director, Task Force on National Homeland Security

We gather here on December 7, 2012, when the solar maximum has commenced. Every 11 years the thermonuclear cauldron that is the Sun enters a phase of heightened activity, hurling into space greater numbers of solar flares and coronal mass ejections that could cause a geomagnetic super-storm on Earth, with catastrophic consequences for the inhabitants of our planet. Most scientists estimate that a super-storm like the 1859 Carrington Event occurs about once every century. If so, we are now 53 years overdue for another Carrington, which would collapse electric grids and the critical infrastructures that sustain modern civilization and the lives of billions worldwide.

Our luck may hold through the solar maximum, which shall last a year, through 2013. No Carrington Event or lesser natural electromagnetic pulse (EMP) catastrophe may visit us, not just yet. But our survival will be due to luck, not to prudence and wise policy. We are living on borrowed time.

Reports are surfacing from sources inside the Iranian Revolutionary Guard that there are two more previously unknown underground facilities working on nuclear weapons, that the program is more advanced than previously estimated, that Iran is now in the process of actually building nuclear warheads. If these reports are even partially true, within months Iran may be capable of executing a nuclear EMP attack against the United States.

When I left the CIA to work as professional staff on the House Armed Services Committee in 1995, one of my first jobs was to help Congressman Roscoe Bartlett educate policymakers and the public on the EMP threat, and on the necessity of protecting our nation. To me, the threat was so clear, the consequences so grave, and the necessity of action so urgent, that I thought we could pass a bill in a year to protect the nation from EMP, and then move on to other national security issues.

Thirteen years later, at the end of 2008, when the EMP Commission delivered its final report and recommendations to the Congress, I thought there was still time to protect the nation. The solar maximum was still four years distant. Iran had not yet developed nuclear weapons. Surely, I thought, the Congress will now enact legislation to protect the critical infrastructures from EMP. After all, historically, Congress has promptly implemented the recommendations of its other Congressional Commissions, or so I told myself in 2008.

Today in 2012, four years after the EMP Commission presented its plan to Congress, 17 years since embarking on the crusade to make America safe from EMP, we are still naked. I have grown old witnessing Washington's ignorance and inertia conspire to make the people vulnerable to the greatest threat to our civilization, yet a threat that is easiest and least costly of solution.

As of December 2012, the national electric grid and other critical infrastructures are still unprotected from EMP. Not a single recommendation of the Congressional EMP Commission to physically harden the critical infrastructures has been implemented.

Yet there is still hope. There may be time to act decisively to protect the critical infrastructures before nature or man hurls the EMP hammer. Given good planning, adequate resources, and the will to move quickly, much could be accomplished even in one year. For example, the 300 most important EHV transformers servicing the most populous cities, and the nuclear power reactors, might be protected on an accelerated basis.

And the year 2012 has seen some amazing progress toward national EMP preparedness, even if it is as yet only "paper progress":

President Obama's Presidential Decision Directive 8 has, for the first time, led to the recognition, in the Strategic National Risk Assessment developed by direction of the White House, of natural EMP from a geomagnetic super-storm as one of the greatest threats for which the nation must prepare. If President Obama invested some of his political leadership in the cause of EMP preparedness, if he gave EMP protection the kind of financial and regulatory support that he accords to climate change, the White House could get the job done, and perhaps single-handedly save the nation.

President Obama's re-election could well be fortuitous for the cause of EMP preparedness. President Obama already knows about the EMP threat, so he will not have to learn on the job. His inherent skepticism about the claims of industry, and his greater trust in government expertise, would serve well the cause of EMP preparedness, in our struggle with industry lobbyists and the NERC. President Obama reportedly will try another round of government spending to stimulate the economy, and he has repeatedly promised investment in infrastructure projects. Any money spent on protecting the critical infrastructures from EMP would be money well spent. As a second term, President Obama can and probably will act boldly, to the very limits of his authority, to achieve his objectives. Such a President is needed to rescue the American people from the looming EMP catastrophe.

Congress in 2012 launched numerous initiatives designed to greatly advance civil–military preparedness for EMP. At minimum, these raised consciousness and greater conscientiousness in the departments and agencies of the U.S. government.

The Department of Defense in 2012 continued to harden military systems against EMP. Although these hardening programs are intended for war fighting purposes, they inherently make the DOD better able to help the Department of Homeland Security recover the critical infrastructures and rescue the American people in the aftermath of an EMP event. During Hurricane Sandy, the Defense Department cooperated and made available DOD resources to prepare and recover from the hurricane to an unprecedented degree. Hopefully this is a new trend that will lead to the much greater and more ambitious cooperation needed between DOD and DHS to achieve the necessary civil–military preparedness for an EMP catastrophe.

The Department of Homeland Security in 2012, for the first time in a congressional hearing, acknowledged its responsibility to protect the critical infrastructures from a natural or nuclear EMP event. As 2012 draws to a close, the DHS is finally working on and may adopt a new National Planning Scenario focused on EMP—one of the chief recommendations of the

Congressional EMP Commission in 2008. All emergency preparedness planning, training, and resource allocation at the federal, state, and local levels is based on the National Planning Scenarios. The inclusion of a National Planning Scenario for EMP would be an enormous step forward.

In November 2012, the U.S. Federal Energy Regulatory Commission directed the electric power industry to develop a plan to protect the national electric grid from natural EMP—including by hardening the grid. Technical hardening of the grid from natural EMP will also provide some, though incomplete, protection against nuclear EMP.

In 2012, despite amazing progress toward EMP preparedness in the policy realm, there were also severe defeats and setbacks:

The SHIELD Act, the single most important congressional initiative for EMP preparedness, that would mandate protection of the national electric grid instead of relying on the voluntary cooperation of industry, failed to pass. Industry lobbyists managed to keep SHIELD bottled-up in the House Energy and Commerce Committee. Congressman Trent Franks will reintroduce the SHIELD Act and fight for its passage in the next Congress.

But now SHIELD cannot pass in time to be of any use protecting the grid during the solar maximum, and perhaps not before Iran gets the bomb and the capability to make a nuclear EMP attack.

Worse still, the champions of EMP preparedness in the Senate are retiring, and will not return in 2013. Senators John Kyl, Joe Lieberman, and Jim Webb will be gone in 2013, with no one to take their place on the firing line, at least not yet.

The bad economy is crippling the EMP movement at the grassroots level. People who fear for their jobs and the immediate welfare of their families must be less generous donating their time and resources for political activism. Even my own Task Force on National and Homeland Security, which has operated on a voluntary basis with very little financial resources through 2012, may perish in 2013.

Worse still, Congressman Roscoe Bartlett—who started the EMP movement and has long been the national leader on EMP preparedness—lost his seat in the 2012 elections. Maryland Democrats gerrymandered Bartlett's seat, making his re-election impossible.

Even though the EMP movement is on the brink of victory in the crucial campaign for sound government policies, our losses in people may well cost us the war. The EMP crusaders have never been numerous or well funded. EMP preparedness has been a David versus Goliath struggle pitting a handful of political leaders, scientists, and grassroots activists, having virtually no resources, against the twin Goliaths of government bureaucracy and industry having virtually unlimited monies and armies of lobbyists.

In 2013, we EMP crusaders may come to our Alamo. I will hope instead for our Thermopylae, or better yet a Salamis.

And I hope fervently, fighting against fear and resignation, that our struggle for national preparedness is not terminated by the sun or a nuclear adversary, delivering upon us an EMP Apocalypse.

Dr. William Graham, President Reagan's Science Advisor and Chairman of the EMP Commission, had a plaque in his office with a quote from Winston Churchill: "Never, never, never, never give up!"

The man I will introduce next, Congressman Roscoe Bartlett, is the very embodiment of those words. Perhaps his best introduction is from a tribute recently given to Congressman Bartlett by Dr. Graham and Ambassador James Woolsey, former Director of the Central Intelligence Agency. Their tribute reads:

CONGRESSMAN ROSCOE G. BARTLETT
IN RECOGNITION OF HEROIC LEADERSHIP

Laboring virtually alone for years,
Congressman Roscoe Bartlett
warned and educated
Presidents and the American People
about the catastrophic threat from
electromagnetic pulse (EMP),
held the first unclassified hearings on EMP,
established the EMP Commission,
and inspired a national movement for
EMP preparedness to safeguard
American Civilization
and the lives of millions.

And we gathering here today are yet another tribute to Congressman Bartlett. He is responsible for every one of us being here. Alone in the Congress, where no one had ever heard of EMP, Congressman Bartlett began his lonely crusade, which has flourished into a growing national awareness of this gravest and most imminent of all threats.

Ladies and gentleman, Congressman Roscoe G. Bartlett!

TRANSCRIPT:
View of the Past and Future Role of Congress in Addressing EMP, GMD, and High-impact
Threats to Critical Infrastructure
Congressman Roscoe Bartlett

Dr. Pry and Congressman Roscoe Bartlett*: Dr. Peter Pry, formerly a CIA analyst and staff to Congressman Bartlett and the EMP Commission he helped launch, reviewed the history of those trying to address this issue beginning with the formation of the EMP Commission to current day attempts at legislation to address the issues. He introduced Congressman Roscoe Bartlett who noted that just as people think some things are "just too good to be true", that similarly, "other things are just too bad to be true," claiming that EMP and these high-impact threats are examples of the latter. Congressman Bartlett's long-term history of wrestling with high-impact threats to infrastructure such as the electric power grids parallels his long standing congressional leadership in energy security and the role of renewable energy. Congressman Bartlett reviewed the history of his involvement of both high-impact threats to critical infrastructure with special emphasis on manmade EMP and space weather along with his related interest in energy security. He answered audience questions and led questioning of other presenters throughout the morning session.*

CONGRESSMAN ROSCOE BARTLETT: Peter is way, way too kind. Thank you very much, and thank you all very much for being here. The other evening we were coming out of Washington. It was a beautiful, clear evening and the lights in all of those buildings were on and reflecting across the Potomac. And I was thinking about this subject as we drove out. Those lights were on, of course, because of electricity. I watched the airplanes coming down the Potomac one about every minute to land at the airport, and the runway lights were on because of electricity and the air traffic control center was functioning because of electricity, or they wouldn't be able to land.

In one of those buildings was a couple going to dinner. They had stopped at the ATM machine to get some money. That wouldn't work without electricity. And they had direct deposits of their paycheck in the bank, and that of course, that whole system wouldn't work without electricity. They took the elevator to the restaurant floor, and of course that was powered with electricity. They ordered their meal and the food was brought to the hotel by a truck, and it was electricity that powered the pumps that put diesel in the truck.

It was also electricity in a large number of ways that was important in the production of the food that they were to eat. They had to go to the restroom during the meal and it was electricity that powered the pumps that brought the water to flush the toilet. And it was electricity that was powering the sewage treatment plant to treat the sewage.

When I thought of all of the contributions that electricity was making to our life there are about 17 infrastructures out there, important infrastructures and none of them work without electricity. Without electricity everything stops. And there are several things that could disrupt the grid perhaps for long periods of time. One of those I think is unlikely and that is a pandemic where everybody is so sick that they just can't get there to keep the system working. Let's hope that doesn't happen.

Dr. Pry has mentioned two of the threats to the grid. One of them is a maybe, and that is an EMP attack. The other is an absolute certainty. That's a when, that's not an *if*, and that's a giant solar storm. The 1859 storm described by Dr. Carrington from England called a Carrington Event that will occur again, and when it does if we have not prepared the grid will be down and it will be down for a long period of time. There was a storm not quite so large in 1921. When we have another one of those the grid will also go down. So this is a when; this is not an *if*.

There are two other things that could bring the grid down. One is a terrorist attack. And I'm told that there are 11 critical substations, so 11 people armed with nothing more than a 22 and the knowledge of which insulators to take out could bring down the grid. The fifth thing that could bring down the grid, of course, and there have been a plethora of articles on this—I get clips every day and a day does not go by where there is not a clip in there mentioning the vulnerability of our grid to a cyber attack.

Now how do these things bring the grid down for such a long period of time? It would seem that after disruption you would just turn the switches and it would start again, but it can't, because almost all of these would causes surges of electricity which would blow the major transformers, and we don't make any of them. There are a few spares, nowhere near enough, to replace the hundred or 200 that experts believe would be taken down in any of these catastrophes. We don't make them. You have to order them. There are none on the shelf anywhere. They will build them when you order it, and it takes a year, a year-and-a-half or so to build them.

The top security person in FERC sat in my office and told me that the grid would be down for 18 years to two months after one of these attacks. I said, "Gee, what would be the consequence?" He said, "Two-thirds of the people in our country will die." That may be an underestimate. There have been a couple of books written on this subject. One I hope you've read is *One Second After*. Bill Forstchen did a really good job of research of that. I think that that is pretty true to what would happen.

Another book, and I came to my office, and I've never met the author of this book, Dr. Lowry, and he was in his hotel—I'm sorry, he was in the hospital room recovering from heart surgery and he was surfing the television, and he happened on C-SPAN. And I was giving one of the probably half dozen hour-long talks I have given on the floor of the Congress about EMP, and he got turned on by this and he did a lot of research. He was a PhD in electrical engineering so it was right down his alley, and he ended up writing a novel. I never thought I could read a 700 page novel, but I read it and wow. *One Second After* is bad enough. What Dr. Lowry described in his book is just absolutely, absolutely horrendous.

Being so dependent on this infrastructure, it is quite remarkable that we have no fallback position. What would we do? What would we do if this happened? Well, this is a fulfillment of a dream to see this many people for this long to come apart to talk about this subject, because it really, really is critical to our future.

I ran for Congress 20 years ago because I didn't have quite so many grandchildren then. I had 10 children. Now I have 18 grandchildren and two great-grandchildren. I was just concerned that they weren't going to have the chances that I had, a really poor Depression era kid, to work and achieve the first person in my direct family to go to college. I got a Master's and a Doctorate and was able to achieve in a number of different areas, and I was concerned that my kids and my grandkids weren't going to have that opportunity. And I am concerned that the basic fabric of the country might not even be there for them if we haven't prepared.

And this preparation is very difficult, because unless everybody does it nobody is going to do it. If one of the power producers decides to do this, to harden their part of the grid and so forth their product is going to cost them more money and they can't compete in a marketplace which is very competitive, and so they're not going to do it unless everybody does it. I'm not a fan of big government, but I suspect this is one of the places where government needs to get involved or it's not going to happen.

Thank you all very much for coming today. Do everything you can to push this ball forward. I hope it's not too late. It could be, but we can't turn back the hands of time. All we can do is the best that we can do from now on. Thank you very much for coming today.

CHUCK: Is Drew Nishiyama here by any chance? I didn't see him in the audience yet. Okay, he was hoping to be here today. If you maybe have a quick one moment I think we have a roaming mic. Does this work? I'm not sure if that does or not. No. Okay, I'll try another one. I know it works somehow. It's just user incompetence. There is nothing wrong with that equipment. How about this?

So, in a moment what we're going to try to do throughout the course of the day as time permits is when a speaker presents we'll have somebody with a prepared question who will be able to immediately ask the question and then be able to have a moment or two to have a couple of questions from the audience. We're running short of time, because we got a little bit of a late start, so what I'm going to do is I am going to pose a question that I knew Congressman Frank's office was going to ask the Congressman. So I'll ask the question briefly and he'll respond however he would like, and then we'll get ready for our next speaker.

So, the question that he would have posed today was, "Given all of your experience in Congress over these last number of years and how difficult it's been to get people to understand these really tough issues, what kind of recommendation would you give the rest of us here as we think about trying to move the ball forward down that field?" And I would be very interested in just getting your words of wisdom to the rest of us as we try to move this ball forward in the next year or two.

BARTLETT: Thank you very much. The innocence and ignorance on the part of our general population about this subject is astounding, and we have truly representative government. So the biggest challenge that we have is to educate government. They would like not to be educated, because this is very uncomfortable to think about.

When I first started thinking about EMP I called my friend Tom Clancy who had had an EMP scenario in one of his books, and I know he did great research and he could tell me something about EMP. He said, "If you read my book you know all I know, but let me refer you to the smartest person hired by the U.S. government." That's a lot of people and a pretty steep hill to climb to be the best one of those, but in his view it was a Dr. Lowell Wood from Lawrence Livermore. And Lowell said, "It's just too hard. They don't want to think about it." And I think that's true. It is really hard. It is really tough.

But if you don't think about it any one of these five things that we have talked about could end life as we know it. Now when you tell somebody about that some things are too good to be true and the corollary of that is that some things may be just too bad to be true. That's just not so. Any one of these five things could bring down the grid, and if you think about it almost everything we do is dependent on electricity in one way or another, everything, and there is no fallback position. There is no fallback position.

There is three days of food in the average city. If you go to Wal-Mart and buy your food there they are 24/7 many of them, they're stocking the shelves as you take the food off the shelves. If you go to Sam's Club they do close the doors there, but they stock the shelves at night. There is no warehouse in back of those stores. The trucks come in to put the food on the shelves while you're taking it off of the shelves.

This is a really critical threat to our society. The biggest challenge that we have is education, just call your Representatives, call your Senators. Tell them this is something they have got to pay attention to. Thank you.

CHUCK: Congressman Bartlett. Now we're going to switch gears in just a little manner in that we're going to be focusing now on one of the other venues, threat venues that could create a long-term collapse of critical infrastructure nationwide in the country, and that is going to be brought by—Yes, we have.

MALE: I have a very quick question. One of the problems is that the public is not familiar with what would happen if we had an EMP event. So I'm asking how would an EMP event compare to the Fukushima event which people are much more familiar with?

BARTLETT: A nuclear bomb dropped on a large city, even New York or London or Tokyo, takes out that city. An EMP attack, one weapon of the right size detonated 300 miles high over the center of our country, Iowa or Nebraska, could produce an EMP about 50 kilovolts per meter. The Russian Generals told us they had weapons that would produce—I mean 100—weapons that would produce 200 kilovolts per meter. At the margins that would be 100 kilovolts per meter. To my knowledge we have never prepared for, designed, or prepared for anything even in our military more than half that.

And beginning with the Clinton years and the diminution of funds in defense we started waiving EMP hardening of our new weapons systems. And I kept asking them on Armed Services, "Why are you building these systems? We don't need those for Iraq and Iran. The only time you would need those systems is against a peer." And in all of their open literature, in all of their war games one of the first things they do is a robust EMP laydown to deny you the use of all of your equipment which is not EMP hardened, and that was essentially all of our new equipment. Almost none of it was EMP hardened.

Now I'm really concerned. I'm really concerned that if the lights go out our military is really not going to be there, because our military now has been de-vesting themselves of all of their infrastructure. The local military base gets their electricity from the grid. It's the grid that powers the water supply that they get. It's the grid that treats their sewage, that powers the treatment of their sewage. I want our military facilities to be self-sufficient. I'm a big proponent of small modular nuclear reactors. Even if there is not irreparable damage to our grid we may not be able to start it, because much of the grid cannot do a black start. They've got to have power to spin up. And if our military bases, which are widely distributed across the country, could keep the lights on during an event like this and have some power to spare for those parts of the grid which were not irreparably damaged they might be able to spin that up again.

Now there are just lots of things that we need to be doing. We should have started it a long number of years ago. And this is a very tough time to do it when we're in a recession and money is very short, but it can't cost too much. If you don't do it it's going to end life as we know it if one of these events happens, and that is what will happen if we do not prepare ourselves.

CHUCK: Thank you very much, Congressman Bartlett. I should mention in light of that question that we have with us later today Dr. George Baker who ran all the technical programs for the government in this area under what was called then the Defense Nuclear Agency. And he is going to be talking about EMP myths and we're going to have a chance to ask him questions. And similarly you'll be able to engage us after the meetings today. We'll put a conference proceedings together, and through the InfraGard EMP SIG, of course, we'll be able to engage you further and make certain that we put you in touch with people who can answer the questions that are inevitably going to be raised during the course of the day that we won't have just enough time to answer today.

So the shift we're about to make now is focusing on a similar but different threat venue which is that of a cyber attack. As you know, the government, including the FBI has a large concern about the issue of cyber. And to address it today we have the leader of that effort from the FBI here with us. But what I'm going to do is have Special Agent Miller come up right now, along with Mr. Richard McFeely who is the FBI Executive Assistant Director for Criminal Cyber Response and Services Branch.

And I just want to say one thing. As you may know, InfraGard wouldn't be able to work well without the generous involvement of the FBI supporting the local chapters and national activities such as this. And when you get to begin to know folks at the FBI they have interesting titles, and many of them have the word Special in it, like Special Agent, and that's because they're special. There must be some technical meaning too, but I'm sure it's secondary to the fact that they're just special people. And very often they'll have other names that go with it, like Really Special Agent so-and-so. So I'm going to call on Really Special Agent Miller to come this morning and tell why our speaker is as special as he is. Thank you.

InfraGard and the FBI Cyber Initiative
Richard A. McFeely, Executive Assistant Director, Criminal, Cyber, Response and Services Branch

EAD McFeely spoke about how the Federal Bureau of Investigation (FBI) and the private sector are working together to protect U.S. critical infrastructure, particularly from cyber threats. The FBI is looking to get the major stakeholders in our critical infrastructure incorporated into the InfraGard model, he said.

EAD McFeely said he strongly supports InfraGard and what the collective power of the association brings to bear. The challenge, he said, is to make InfraGard relevant to its members and a value-added resource for member companies.

He spoke about the threat to the North American Electric Power Grid, noting there is a general consensus that the Grid is vulnerable to multiple physical and/or cyber attacks that would result in significant and sustained loss of power. The FBI is concerned about these vulnerabilities, especially when it comes to attacks against SCADA (supervisory control and data acquisition) systems, global positioning system (GPS)-dependent facilities, and disruption of transmission lines.

While he said the threat of electromagnetic pulse (EMP) attacks should be taken seriously, EAD McFeely focused his remarks on attacks against critical infrastructure from a cyber perspective, which many experts believe represents a higher probability risk.

He noted that FBI Director Robert Mueller has openly said what most USG professionals have believed for a long time; that is, he expects the cyber threat to surpass the terrorism threat to our nation.

EAD McFeely said the FBI recognized the significance of the cyber threat a decade ago and in response:
- created the Cyber Division;
- elevated the cyber threat as our number three national priority (only after counterterrorism and counterintelligence);
- significantly increased our hiring of technically trained agents, analysts, and forensic specialists; and
- expanded our partnerships with law enforcement, private industry, and academia, through initiatives like InfraGard.

He said that by far, the creation of the InfraGard program in the mid-1990s represents the premier example of private–public partnerships. Since then, he noted, we've seen this initiative grow from a great idea in one field office to more than 53,000 members in 86 chapters across the United States. He thanked InfraGard members for the time and effort they voluntarily put into this endeavor. The FBI plans to build on this partnership and take it to new heights.

Over the past year, EAD McFeely said, the FBI—under its legal authorities and in conjunction with our government partners—has had a fair amount of success in warning some potential victims ahead of time that Computer Network Exploitation or Computer Network Attack were about to happen. They were able to use that information to shore up their defenses. The FBI wants to be able to do more of that.

The need to prevent attacks before they occur, he said, is why the FBI is redoubling its efforts to strengthen its cyber capabilities in the same way it strengthened its intelligence and national security capabilities in the wake of the September 11 attacks.

The FBI's Next Generation Cyber Initiative entails a wide range of measures: From focusing its Cyber Division on intrusions, to hiring additional computer scientists, to expanding partnerships and collaboration at the National Cyber Investigative Joint Task Force (NCIJTF).

The part of Next Gen Cyber that is most relevant to InfraGard is the FBI's plan to significantly expand its partnerships with the private sector, EAD McFeely said. That includes enhancing the amount of information we exchange with industry and providing tools to help repel intruders, including a malware repository and analysis tool and an electronic means for industry to report intrusions real-time. In the FBI's cyber strategy plan, Infragard will become one of the centerpieces of this exchange.

To provide a more targeted threat picture and a more meaningful dialogue, the FBI is also proposing to create Sector Chiefs within InfraGard, he said. These individuals would serve as Points of Contact for investigations and issues related to their sectors. They would serve as an expert resource for FBI field offices and other InfraGard members and facilitate intelligence dissemination within their sectors.

EAD McFeely urged the private sector to report intrusions and assured them that the FBI is very careful about protecting proprietary and other confidential information. He asked Companies who are not already a member of InfraGard to join and those who are already members to work with the FBI to organize chapters into sectors so the Bureau can tailor its messages along industry lines.

He concluded by saying that partnerships are critical to the FBI success in all areas, but that is particularly true in the cyber realm because of the constantly evolving nature of technology and the fact that it affects all sectors of the U.S. economy and government. The private sector and InfraGard in particular are a critical part of the fight against the cyber threat, he said, but the partnership must be energized by increasing the back-and-forth flow of information. By giving industry a secure means to report intrusions, an opportunity to have malware analyzed, and other tools and information, the FBI believes it is on its way to doing just that.

InfraGard and the FBI Cyber Initiative
Richard McFeely

Mr. Richard McFeely, *FBI Executive Assistant Director, Criminal, Cyber, Response and Services Branch: Mr. McFeely used his address as an opportunity to not only highlight the growing concern over cyber-security, but, announced a new program that the FBI is launching to work with the private sector to protect their networks using InfraGard as a major program vehicle. The announcement included the proposed use of automated tools to assist the private sector in learning about and reporting detrimental cyber incidents.*

DAVE MILLER: Good morning. As Chuck mentioned I'm Dave Miller. I'm a Supervisory Special Agent with the Federal Bureau of Investigation (FBI)'s Baltimore office just up the road. And it is a great honor for me to introduce today Executive Assistant Director Richard A. McFeely. Mr. McFeely is the Executive Assistant Director of the FBI's Criminal Cyber and Crisis Response Services Branch. He was appointed to this position back in July and from what I can figure he has been a very busy man since he got to headquarters. He is responsible for all FBI criminal and cyber investigations worldwide as well as international operations and critical incident response.

Mr. McFeely entered on duty with the FBI as a Special Agent in February 1990 and was assigned to the Buffalo division and he worked primarily violent crime matters. He was the lead case agent in the Oklahoma City bombing back in 1995 and a few years later was promoted to Supervisor at FBI Headquarters in the Criminal Investigative Division. During his stint at headquarters he oversaw the FBI's drug investigations and about a year after arriving there he was detailed to the Executive Office of the President, Office of National Drug Control Policy where he helped formulate national policy on the nation's counter narcotics efforts.

He transferred to the Washington field office in 1999 and following the 9/11 attack on the Pentagon he was the FBI's on scene commander. He supervised a counterterrorism squad after that and was instrumental in setting up a joint intelligence center in Fairfax, Virginia with state and local police as well as the FBI and other agencies to promote and increase information sharing among the federal, state, and local governments.

He received the FBI Director's Award in 2005 for outstanding counterterrorism investigation for supervising a multinational investigation into the assassination plot against a foreign head of state. He later became an Assistant Special Agent in charge of the technical programs of the FBI.

In 2006 he returned to FBI Headquarters as a Section Chief and became the FBI's Budget Officer and was later the Deputy Chief Financial Officer for the entire FBI, oversaw the entire budget acquisition and accounting functions. Then he moved to Baltimore where he was the Special Agent in charge for the last three years until he came back to FBI Headquarters in his current role.

I had a great opportunity to work with Rick for the last few years. He is one of the best agents I've ever worked with or worked for and it is my distinct honor today to introduce Executive Assistant Director Richard A. McFeely.

RICHARD A. MCFEELY: Thank you, Dave. Good morning. For some time now my wife has been working on a mini electromagnetic pulse (EMP) device to try to get her kids to talk to us again. But I do want to thank you for the opportunity to come to talk with you about how the FBI and the private sector are working together to protect this critical infrastructure particularly from the cyber threat. I was asked to speak specifically about how InfraGard fits into the role of protecting infrastructure security across the private sector, but I want to relate this to how we're looking to get the major stakeholders in our critical infrastructure incorporated into the InfraGard model.

I recognize that not everybody here today is an InfraGard member. I hope after today you go back if you represent a company, if you represent somebody in the critical infrastructure world, that you rethink whether or not the InfraGard model is something that you want to participate in. I'm going to give you a lot of reasons today why.

There has been major transformation going on right now within the FBI, and as you hear me speak today I'm going to announce some things that really have been in concept now over the past six months, but they're going to radically transform the FBI's relationship with the private sector and InfraGard itself is going to be the key component behind all of that.

The challenge that I have seen over the years is how to make InfraGard relevant to its members and a value-added resource to your company. I'm not sure in all honesty that the FBI has really lived up to those tenants. That situation right now is changing at lightning speed, and I want to talk about those changes but briefly talk about the importance of it in context of the current threat. The threat to the North America power grid has been studied internally within the U.S. government, by industry, by the private sector, academia, and think tanks. There is a generous consensus that the grid is very vulnerable to multiple physical and/or cyber attacks that would result in significant and sustained loss of power.

We in the FBI are very concerned that these vulnerabilities exist, especially when it comes to attacks against data systems, our dependence upon GPS, and the disruption of transmission lines. The subject of EMP attacks needs to be studied more and the threat needs to be taken very seriously, but I want to confine my remarks today towards attacks against this critical infrastructure from a cyber perspective, and which to many experts represents a higher probability risk.

One of the significant reasons behind such an assessment, as we all know, is that cyber attacks can be conducted remotely, cheaply, and with off the shelf products available for sale on the internet. We see that every day. I know this audience is all too familiar with the nature of the cyber threat so I won't take our time by describing it in detail, but my reference to changing the FBI's engagement with the private sector is specifically because of the current threat environment. It's no secret that over the past few months that key segments of our financial sector have been under cyber attack. The implications of taking down these key segments of our financial infrastructure are obvious.

Suffice it to say when you look across the world as a whole we are losing data, money, ideas, and innovation to our many cyber adversaries. And our critical infrastructure is vulnerable to attack by those who wish to do us harm. We are a wired world and there is no turning back from that. FBI Director Robert Muller has openly stated over the past few months what most U.S. government professionals believe. That is that he expects the cyber threat to surpass the terrorism threat to our nation.

So what is the FBI doing about that? We recognized the significance of the cyber threat several years ago, more than a decade ago. We created a Cyber division. We have elevated the cyber threat as our number three national priority behind counterterrorism and behind our responsibility for counterintelligence. We have significantly increased our hiring of technically trained agents, analysts, and forensic specialists, and we have expanded our partnerships with law enforcement, private industry, and academia, and through initiatives like InfraGard.

By far the creation of the InfraGard program in the mid-1990s represents the premiere example of private public partnerships. Since then we have seen this initiative grow from a great idea in one field office in the Midwest to more than 53,000 members in 86 chapters across the United States. On behalf of the FBI I want to thank all of you who do participate in InfraGard and who put a lot of effort like today's conference into this endeavor. And I want to thank you for your continued support as we build on this partnership and take it to new heights.

There is virtually no discussion that takes place within our halls when it comes to the cyber threat and private engagement that doesn't bring up the role of InfraGard. Our plan is to build much of our plan capabilities into the InfraGard structure, and in a moment I will talk more about those capabilities.

While we have made significant progress against the cyber threat in recent years it continues to evolve and expand. What you may not be aware of is that in many instances we have insight and advanced notice that adversaries are going to launch an attack. Over the past year under our legal authorities and in conjunction with our government partners we have had a fair amount of success in warning companies ahead of time that they are about to be the victims of a computer network exploitation or an attack, and they were able to use that information to shore up their defenses. We want to be able to do more of that.

We need to prevent attacks before they occur, and that is why we and the FBI are redoubling our efforts to strengthen our cyber capabilities in the same way we strengthened our intelligence and national security capabilities in the wake of the attacks in 2001. We call it our Next Generation Cyber Initiative, and it entails a wide range of initiatives from focusing our Cyber division now strictly on intrusions to hiring additional computer scientists across the country, and in expanding our partnerships and collaboration at the National Cyber Investigative Joint Taskforce, the NCIJTF, which I'll speak to in a moment.

But part of our Next Gen Cyber that is more relevant to you is our plan to significantly expand our partnership with the private sector. In the past industry has provided us information about attacks that have occurred and we have come in in a reactive mode and investigated them. Our adversaries have taken advantage of the fact that we have been limited in the kind of information we have exchanged with the private sector. But we all now realize that we can no longer keep this a one-way communication. We are going to give you tools, including information to help to repel these intruders.

In fact, just two weeks ago for the first time the FBI started to hold classified threat briefings to the major internet service providers to educate them and work with them to exchange information. That is now part of a cycle that every week we are bringing in major ISPs and our backbone telecommunications providers and actually briefing them on the same material that we're seeing that is emanating against our critical infrastructures.

And I want to tell you about some additional specifics, how we intend to engage with the private sector. We are developing a malware repository and analysis tool—we call it BACSS, the Binary Analysis Characterization and Storage System—to help us identify our cyber adversaries and prevent attacks. If you've been hacked you can send the malware into our system and get a report back on whether other companies have had similar attacks. If there is a match the FBI would make contact with those companies to see if they wouldn't mind sharing that information with you.

We want to do that with the FBI as an intermediary so there is no concern with privacy issues. We plan to make BACSS the nation's repository for malware and viruses, much the same way that we do it with fingerprints, with DNA, and criminal arrest records for U.S. law enforcement agencies and our foreign partners.

We are also going to provide an electronic means for you to report intrusions real time. Right now to report an intrusion to the FBI it's basically based on whether the company has a contact within the FBI or somebody in the company picking up the phone and calling the local FBI field office. That is not a good way to do business. We can't treat this as an ad-hoc responsibility. The system that we are developing which will be deployed sometime over the next few months is called I-Guardian and it's modeled after the Guardian system that we have used very successfully for the past 10 years for law enforcement to report terrorism leads to us. It is based off the same platform that we use to track these terrorism leads, and it, as I said, is a highly successful effort on the CT side, the counterterrorism side that we believe now can be rolled out to the private sector and provide an online platform to report intrusions to us.

We also want to ensure that if you do provide us information that we're going to keep it safe. We're currently in the process of vetting a template for nondisclosure agreements through our Office of General Counsel. We make our living protecting proprietary information, but we want to be able to put it in writing to assure you that the reporting you make to us will be kept safe from outside disclosure absent extraordinary circumstances. We recognize that you have to have that degree of comfort from us before you're willing to report the intrusions.

Through InfraGard we're doing better on exchanging information at a higher level about what we are seeing, but it's hard to stand up here and deliver the same message to retailors, to defense contractors, and to energy companies. The threat to each of these sectors is not always the same. To provide a more targeted threat picture and a more meaningful dialogue we are proposing to create Sector Chiefs within InfraGard. These sectors are going to be divided up along DHS's 18 critical infrastructures, and these individuals would serve as points of contact for investigations and issues related to their sectors. They would serve as an expert resource for FBI field office and other InfraGard members, and they would facilitate intelligence dissemination within their sectors.

Concentrating on specific individual sectors allows us to develop a more focused picture as to who wants what from your networks. A country with a weak military, for example, is likely after technology designs, where the motive of those with a more developed economy is more likely using espionage to develop ways to defeat our advanced technology. We want to make sure that we have the private sector as well as U.S. government partners poised to handle that threat.

We are also expanding our cyber training for the private sector. Early next year we're launching the first session of our National Cyber Executive Institute. This is a three-day seminar to train leading industry executives on cyber threat awareness and information sharing. Held in Quantico, our first few classes will be targeting those InfraGard Sector Chiefs that I just mentioned. During this three day seminar you will hear from various U.S. government agencies who we work with very closely, and we'll walk through real life scenarios to help us both develop ways to better defeat the threat.

So what can the private sector do as part of this process? Number one, report intrusions. It is critical that when your network is breached you report the intrusion to us. I am continually amazed every day at the number of intrusions that we're aware of based on our authorities to collect intelligence with our partners that we have knowledge of that we know the companies know about, but the companies never come forward to us. And we go out and we make victim notifications to those companies and, like I say, many times they are aware that they have been hacked, but they have taken no steps to report that intrusion to us or to DHS or anybody else that has an interest in the cyber intrusion world.

We recognize that you may be hesitant to disclose the breach out of concern that the word is going to leak out to the public or to the shareholders, but it's important that you recognize that you need to be part of this solution. We can work together to resolve the type of exploitation or attack you are undergoing. And as I mentioned, we are working on ways to insure the confidentiality of our relationship.

Number two, join InfraGard. If your company is not already a member of InfraGard please look into it and decide whether or not your company belongs and what the value added will be for you, especially those of you that are involved in the critical infrastructure sectors.

Number three, work to expand your existing chapter. If you're a member of InfraGard we want to work with you to organize your chapter into those sectors that I talked about. This is a new concept for InfraGard. We have over the past 10 years basically delivered one unified message to all the InfraGard members. As I explained, that's not a good model. Each of you that comes from different sectors has different issues, has different threats. The adversaries are interested in specific types of information and it's not necessarily a unified message. By breaking us up into sectors we feel that we're going to be able to deliver a much more narrowly focused threat briefing to you and work with you to understand the issues that you might be having.

We need your help in bringing new industry partners. The great thing about InfraGard is these things, like BACSS and I-Guardian that I spoke about, they're going to be available through the InfraGard portals. In other words, we're going to be using InfraGard as the hub to transmit that information directly into the local FBI field office. You need to be full partners with us through the good and the bad. If you think that the FBI or the U.S. government has full-on solutions to these threats you are mistaken. We learn as much from you as you learn from us. Oftentimes you in the private sector are seeing the threats before we're seeing them. We have to make sure that we're linked up. The only people that are winning right now are the bad guys.

As in all of our work these partnerships are critical to our success, but it's particularly true in the cyber realm because of the constantly evolving nature of technology. The bottom line is we need you. You are a critical part of the fight against the cyber threat, but we've got to energize this partnership by increasing the information flow back and forth and by giving you a secure means to report your intrusions, an opportunity to have your malware analyzed in this repository and other tools and information. But we believe through these things, these are great first steps in engaging the private sector.

I want to thank you again for inviting me to be here today.

CHUCK: Before he leaves the stage I'm going to have Special Agent Lauren Schuler ask a directed question. And before she does, I was just reminded in the EMP SIG we're also organizing private sector leaders in various infrastructures and subject matter expertise to become advisors to us on a national basis, and we have a number of people today who have agreed to do that, many of whom are speaking today, folks in subject matter areas such as Dr. Baker; Congressman Bartlett has agreed to do that on the policy side; there are a number of others who will serve as national experts that could be available to local chapters. And so I think that would be a great parallel to the work that you're doing with all of these critical infrastructures as well.

So, what I would like to do is having Lauren ask the first question, and then I think we'll have time for one or two questions more. And Laruen, why don't you begin with the first question.

LAUREN SCHULER: Good morning. Can you hear me okay here? I'm Special Agent Lauren Schuler. I'm the InfraGard Coordinator in the Baltimore field division. Thank you, Mr. McFeely, for being here today and providing these great remarks and for your support of InfraGard. I know you've provided a lot of information this morning on what the FBI is doing to work with the private sector to prevent attacks on our infrastructures. I wondered if you could just take a minute and clarify the FBI's role with the private sector in the event of an actual attack on our power grid or to our cyber infrastructure or to one of the 18 infrastructures that we've been discussing this morning.

MCFEELY: Having been a veteran of the huge reorganization of the United States government after the attacks in 2001 there was a lot of chaos, as everybody knows. The Department of Homeland Security was created. It consolidated dozens and dozens of agencies. There was a scramble, of course, to figure out who the terrorists were that conducted it, what was the extent of their networks. And there was a lot of back and forth between government agencies as to who was responsible for what.

I'm happy to say that one of the best things that is going on now at the highest levels of this government is the discussions among the interagency organizations that have a role in the cyber threat that there is great cooperation. There isn't anything that the FBI is not involved in that doesn't involve our fellow intelligence community partners and with the Department of Homeland Security. This is not a situation that existed right post-2001 where everybody was kind of left to fend for themselves and define their lanes in the road. The lanes in the road as the threats come in are being discussed jointly and with whole scale agreement as to who is responsible for what.

Obviously the scenario you just described is the worst-case scenario—we have a situation where something has already happened. Much of the effort right now in the U.S. government is to try to prevent any such attack. The impact obviously of post-attack is going to involve a whole host of agencies. From the FBI's perspective it's obviously going to be the who and the why. We'll get to the who based on the how. In other words if the infrastructure electric grid was brought down by a malware attack or something like that it's going to be through cyber investigation between us and our national security partners to try to trace it back to who is behind the keyboard.

We spend an awful lot of resource and efforts doing that now, but it's not just an intelligence community or law enforcement response. The consequence management behind this is well in the lane of the Department of Homeland Security who I think is well postured at this point to handle getting the networks back online, to mitigating any potential future damage. So, it is really going to be a wholistic government response and, like I say, this is one of these areas where there is not in-fighting within U.S. government. These types of scenarios are talked about every week in the interagency.

The NCIJTF that I spoke about earlier is 19 agencies that have come together. It's run by the FBI, but the Deputy Directors are Deputy Directors from NSA, from CIA, from the Department of Homeland Security, and we exist really to have not just an operational response but at the policy level to talk about the what ifs. And that is an example of the kind of wholistic government approach that we would bring to bear.

CHUCK: Okay, I see a couple of hands. Peter is close to the—Oh, Congressman Bartlett, he's in the front row. He gets a first shot. He would like to ask a question.

CONGRESSMAN BARTLETT: Thank you very much. If in fact the grid is down, really down for a prolonged period of time, which any one of these five things could produce, how much functional government really remains?

MCFEELY: Well, I hate to stand up here and kind of give the after action report with a bomb damage assessment to our adversaries. Obviously it's a huge concern and we would rely within the FBI on probably the DOD elements to help us. Obviously FBI agents are going to experience the same sort of issues that private citizens are going to experience. They have families. They're going to need to feed their families. We are going to have to maintain essential government services and it's going to be a very difficult road to hoe.

But there are those discussions going on right now. I'm relatively confident the communications can be quickly restored among government agencies, but there is obviously overriding concerns on multiple levels. So, I wish I had a better answer for you, but I think it's one of these issues that need to be studied more carefully.

CHUCK: I think we have time for two quick ones left. Peter?

PETER: I'll make it very quick. I don't actually have a question but a point of information. On the issue of EMP versus cyber, the Congressional EMP Commission found that when you look at the military doctrines of our foreign adversaries—Iran, North Korea, Russia, China—they define cyber warfare differently than we do. They include EMP, nuclear EMP, EMP from radio frequency weapons, as well as computer viruses. If they go after our critical infrastructures it's not going to be one of these things. They're going to throw in the kitchen sink, especially in a war.

And Cynthia Ayers from U.S. Army War College is here, but the U.S. Army War College made a recommendation to Cyber Command that we need to adjust our military cyber doctrine to reflect this reality.

CHUCK: Thank you very much, Peter. And one more question.

MALE: Good day, sir. I have a question about how the cyber archive of incidents, whether you're getting feeds from say Verizon, cyber incident listing CEI cert, U.S. cert, overseas archives of cyber incidents, because right now I'm not seeing that. How are we addressing that?

MCFEELY: You mean whether the FBI is hooked up at that level?

MALE: Affirmative.

MCFEELY: So, I can tell you that we are lockstep in with DHS and through their certs. The Assistant Director for our Cyber division has multiple contacts with the DHS entities responsible for running the certs, not only domestically but overseas also. There are dozens of certs also overseas and there are relationships between the FBI and all of them. The information flow between the FBI and the Department of Homeland Security is at an astounding level of effort, and we're not dealing with any issues right now on exchanging such information.

The lanes in the road, like I say, are fairly well defined. DHS's role is firmly in the prevention, mitigation and recovery end, and we recognize that that is a key piece to any consequence management. And the FBI's role is obviously the investigative role, but we also have a prevention piece that as we're finding malware or finding intent of actors to actually attack networks that we have got to get that information very timely to DHS so they can do what they do and disseminate the information out to the private sector. That's happening every day in multiple, multiple instances. I can assure you that we are lockstep in as the U.S. government between the Department of Justice and Homeland Security.

CHUCK: Thank you very much. These are truly substantial and transformative changes that he is announcing to us for the first time. We are very appreciative of the fact that you would do that, especially for us in this audience who are honored to have that. Thank you for that.

Engaging Utilities for Honest Appraisals of Grid Security
Tom Popik, Chairman, Foundation for Resilient Societies

Good morning, I am very appreciative of the opportunity to speak to you today about electric grid reliability in the United States. I am not an expert on the electric grid nor have I ever worked for an electric utility. In fact, I am a private citizen who lives in southern New Hampshire, an area painfully prone to grid outages due to weather-related causes. Within just the past year, my family went nearly a week without grid power after an ice storm.

In the fall of 2010 I had extra time during a business trip to Washington DC and decided to attend a public meeting of the Federal Energy Regulatory Commission (FERC). I arrived about one hour early and introduced myself to the industry lawyers and lobbyists who also arrived early. To my great surprise, I was told that I was only the second private citizen in recent memory to take an interest in the affairs of FERC regarding electric reliability, the first being a federal prisoner who could not attend the meetings in person.

I introduced myself to Joseph McClellan, then Director of the Office of Electric Reliability at FERC. Mr. McClellan told me that FERC regularly received dozens of docket comments from the electric utility industry, but not a single comment from the public. This made it very difficult for his department to balance the public interest against the corporate interests of electric utilities.

At the time, I knew that thousands of Americans were concerned about the electric grid going down and being left without light, heat, sanitation, telecommunications, police and fire services, and even food and water. I knew this because of my own experience and the work of advocacy groups such as EMPACT America and the Electric Infrastructure Security Council. In September 2010 I had attended a three-day conference at the Army War College appropriately titled, "In the Dark; Military Planning for a Catastrophic Critical Infrastructure Event." And importantly, I had learned about the excellent work of the Congressional Electromagnetic Pulse Commission, which had published two unclassified reports on electromagnetic pulse threats to the U.S. electric grid.

I am happy to report that since 2010 multiple advocacy groups have begun to publicly comment on FERC dockets for electric reliability. Seeing an unmet need for public comment on the connection between electric grid reliability and nuclear safety, I co-founded a nonprofit group, the Foundation for Resilient Societies. We submitted a 100-page Petition for Rulemaking to the Nuclear Regulatory Commission describing the dangers to nuclear power plants were the United States to be hit with a severe solar geomagnetic storm and lose outside power.

Our draft petition was first submitted to the NRC in February 2011, a full month before the disasters at the Fukushima nuclear power plants in Japan. The events at Fukushima amply demonstrated the effects of loss of outside power on nuclear power plants. We asked the Nuclear Regulatory Commission to develop, through rule-making, requirements for long-term backup power for spent fuel pools at nuclear power plants.

Again, I am not an electric reliability expert and receive no compensation from my Foundation for our grid-protection research and education activities. I know many of you in this room are bona fide infrastructure experts and others of you fall into the concerned citizen category like me. All of us can do a great service to our country and to our communities by bringing public and government attention to the vulnerabilities of the electric grid and the widespread non-preparedness of electric utilities for natural and man-made disasters. Further, we must advocate for cost-effective measures to prevent severe blackouts, and to mitigate more effectively the blackouts we cannot prevent.

Our situation is dire indeed. According to a report prepared by the Oak Ridge National Laboratory and sponsored by FERC and the U.S. Department of Energy, an extreme solar storm could leave 100 million Americans without electric power for a period of 1–2 years. A recently declassified report by the National Academy of Sciences disclosed that a terrorist attack on key transformer locations could also produce widespread and long-term electric grid outages. And we know from recent experience that large-scale grid outages are becoming increasingly common, with significant economic and societal consequences. Examples include the February 2011 Texas Cold Snap Outage, the August 2011 Hurricane Irene Outage, the September 2011 Arizona—Southern California Outage, the October 2011 New England Ice Storm Outage, and most recently the Hurricane Sandy Outages in New Jersey, New York City, Long Island, and elsewhere.

The electric utility industry and its self-regulatory body, NERC, have too often not demonstrated a proactive track record in improving electric grid reliability. An excellent case study involves protection against solar storms or so-called geomagnetic disturbances. A moderate solar storm which hit the Canadian province of Québec in March 1989 produced a province-wide blackout and conclusively showed that solar storms can cause grid collapse and permanent equipment damage. In the 23 intervening years, NERC and the electric utility industry have done little to protect against solar storms other than adopt so-called "operating procedures". These "operating procedures" are only marginally effective in protecting critical and hard-to-replace equipment, but instead are designed to prop up the grid with so-called "reactive power reserves." "Operating procedures" depend upon warning from a single satellite that is long past its planned operational life and could fail at any time.

As I travel around up and down the East Coast for my day job, I marvel at our great cities that could go dark for years should this single satellite fail to give warning of an impending "Carrington Event" or, with only brief warning of a severe storm, should our electric utilities fail to de-energize their vulnerable extra high-voltage transformers. And even if large portions of the grid were to be shut down in anticipation of a severe solar storm, black-start recovery procedures are by their very nature untested. With increasing interdependency between natural gas pipelines distribution and electric generation; recovery from any widespread grid collapse will be difficult indeed.

In August of 2011 Dr. George Baker, another director of the Foundation for Resilient Societies, and I became official observers to the NERC Geomagnetic Disturbance Task Force. We have traveled to NERC headquarters in Atlanta on multiple occasions to attend GMD task force meetings. The U.S. Congress, showing extraordinary wisdom, mandated that meetings of NERC committees be open to the public in the legislation establishing NERC as a self-regulatory body. NERC and the electric utility industry have resisted and delayed reliability standards for geomagnetic disturbances and also standards for cyber security, and standards for protection against hurricanes and other storm-related outages. It took five years for NERC to approve rudimentary standards for cyber security—including identification of critical assets—and also five years to approve a standard for so-called "vegetation management" to reduce weather-related outages.

To be fair, we need to recognize that the U.S. electric utility industry is composed of thousands of individual power producers and distributors, with overlapping state and federal regulation. There are competing demands for environmental protection, use of renewable power generation resources, low-cost power production and affordable rates, and electric grid reliability. However, should the United States experience a widespread and long-term grid outage, all of these practical difficulties will be forgotten during the government inquiries and public outrage that will follow.

How can we as infrastructure experts and/or private citizens prevent impending disaster? We need to individually and collectively make our concerns known to regulators and legislators who can force new standards for electric grid reliability. The professional staff and commissioners of the Federal Energy Regulatory Commission (FERC) have been both receptive and proactive in attempting to establish new reliability standards. But unfortunately, existing law only allows for FERC to request or order that NERC and the electric utility industry develop reliability standards—FERC cannot set reliability standards on its own.

Even within the existing regulatory structure, docket comments from the public on proposed electric reliability standards can make a substantial difference. Currently, FERC has an open rulemaking docket on reliability standards to protect against solar storms and resulting geomagnetic disturbance. I ask each of you to consider entering your own comment on the FERC Rulemaking Docket RM12-22 for geomagnetic storm protection. The primary sources of these geomagnetic disturbances are solar storms and man-made electromagnetic pulses from nuclear weapons. During one of the breaks, if you give me your business card, I will provide you background and instructions for submitting a docket comment. The comment deadline is December 23rd.

I also ask you to continue to be involved in organizations that promote critical infrastructure protection, including protection of the electric grid. The FBI InfraGard Special Interest Group on electromagnetic pulse is a prime example of an organization that can have significant influence on governmental processes.

Again, I would like to thank you for the opportunity to tell my own story and to encourage each of you to become involved in the critical cause of electric grid reliability. I look forward to your candid and direct questions.

Lessons from the Fukushima Disaster
Yuki Karakawa

Mr. "Yuki" Karakawa, Chairman of the Japan Resilience Initiative Task Force, provided an in depth review of the nuclear disaster around the recent Fukushima Tsunami and quake. According to Mr. Karakawa, the similar resistance of government officials to fully engage this high-impact issue resulted in unnecessary deaths and continued exposure to toxic levels of radiation from the yet uncapped radioactive plants in Fukushima. He reviewed the need for local communities to be more resilient and for enhanced response capabilities. He emphasized the need for a constitutional change that would make it possible for Japan to issue an emergency declaration. All of this was not lost on the audience who assembled to hear him on this "Pearl Harbor Day".

CHUCK: So now what I would like to do as we get started again is introduce you to Mr. Yuki Karakawa. He's the Chairman of the Japan Resilience Initiative Task Force. It sounds a lot like something Tom might do, right? Except if you look at his bio he has a lot of interesting corporate and government relationships that he has been involved in Japan. And I first met him just this past summer at National Defense University. They host an annual sort of like a conference or demonstration called Start Tides where people are involved in disaster relief or of any kind actually, including non-governmental organizations in this country to show off their technology and what they do. And I was able to meet him there, and I thought it was very appropriate because of our concerns about the fragility of the electric grid and the potential impact on nuclear power plants here.

We have a lot of lessons to learn, I guess, from Japan, and as Tom said they do a lot of things better than we do, and I'm very glad that he was able to come here today. And also, this is a side bar that shows that many of us really care about Japan. We think very highly of them and we welcome them all of the time, but lots of times when we're sitting over here and something happens if you're not involved in the international Red Cross or these folks at NDU who work with these relief agencies you sometimes feel helpless to help people after the fact.

Your being here today can make a difference, because you can do something before the fact. But to show you how helpless you can sometimes feel the day Fukushima happened I hadn't had a watch for some years, because I'm using my cell phone all the time, and I said, "I'm going to go do something for the country of Japan." Little me, what do I have? Nothing that I can do. So I went and bought a Japanese watch from Seiko. I assume it's still a Japanese company and hasn't been bought by China or something. So, that was my attempt to do something, but I'm very glad that I was able to meet Yuki and that he was able to come here, take a few minutes to discuss some of his involvement. And I'm sure he'll be interested in some of your questions about how they do things here versus what we do and the issues involved with making awareness happen and trying to be responsive in some way. So at this time I'd like to invite Mr. Yuki Karakawa to come forward and present us with his background and his presentation. Thank you, Yuki.

YUKI KARAKAWA: Chuck, thank you very much for a nice introduction. I am honored to speak in this Dupont summit. And here most of the people are experts for EMP, right, so I don't touch about neutron and the magnetic field interference and also phased EMP storm system with a geomagnetic storm. I don't touch these issues. I want more touch about politics, what has happened and why Japan is still doing.

And now in Japan disaster still continues. You see today we have a big earthquake in Japan and always everyone is in a shelter, so all work is stopped always or every day work is stopped. And last year March 11th you remember the [00:03:57] earthquake. This is not a direct hit, but most of damage was caused by tsunami, and this was a closed state disaster. So, we have several Governors, so Governor, if we have one Governor only then earthquake it's very simple. We had several, so Governors, every Governor fight each other with Prime Minister. This is another political problem.

And another one is we see the tsunami and though it's just a simple disaster we finished in or after so many years, but Fukushima radiation things are different. It's 150 years project. And this is a map last time we used, and this is by U.S. [00:04:51], U.S. aid and U.S. military responded very well. [00:04:59] U.S. military base and Japanese military did not have much capability. Then we had [00:05:08] project. And earthquake itself, we did not get much direct impact, but the tsunami was bigger. And this is some video. This is a close to Sendai Airport.

So, most of damage is not by earthquake. This was by tsunami. And by the way, in Fukushima [00:05:40] that big damage is not by tsunami; it's by earthquake, because a fault line is just under the plant. In the tsunami evacuation, we have a policy and so, but it did not work much. And this is a Fukushima case. You can see first explosion is the bottom picture, and it just explodes or explodes horizontally, and the second explode is the picture in the middle, it's more horizontal and the black cloud like a weapon, a nuclear weapon.

And the big issue for a nuclear plant is, one of the headaches of an electric company is this Fukushima plant is not insured, so they did not insure the insurance for nuclear accident. This was just it happened. And tsunami just hit the power generator, and they usually have a safety system. It was designed originally, but that safety shutdown system was cut when they constructed 40 years ago. Maybe it cost high then.

And the information is not available to the citizens well, so everybody was outside and gets the sunshine. And most of our children are dosed because of our radiation and also [00:07:16].

And this is a simulation by a meteorological agency or council. This is a schism, where a schism flew. **(?)** [00:07:27]. This yellow is a schism. So it flew about 1,000 kilometers and 300 kilometers should be evacuated, but we did not, Japan did not evacuate. And left side, this map is a heavily damaged area. It's all a schism and the [00:07:53] and the iodide. In some areas we can find very strange materials like a [00:07:59]. Now our defense force is collecting all materials and also [00:08:05] because a second explosion is more like a bomb and explode and it flew like 30 kilometers. So government or defense is now collecting our [00:08:20] pellets in the city.

And you already have a—this is just a summary. You already have knowledge in radiation. Most of Japanese people misunderstood that radiation and the nuclide is different. A nuclide particle will come from a power plant damage and also like uranium mines, and the radiation will come from an x-ray and a nuclear bomb. A nuclear bomb will emit mostly x-ray or heat, so most of our damage in Hiroshima was heat. Heat killed most of people, not by radiation.

And we have a type measure. Nuclide is an iodide and the schism [00:09:20] so it's emitted from our nuclear power plant. And still in Chernobyl the plant was sealed in two weeks, but in Japan it's still open, so every summer or when temperature became high or like a summer always nuclide will fry. And it's raining so cloud is radiated. So, still even in Tokyo when we get rain we should be in a building for like two hours, because cloud will—nuclide is in the cloud. Then that nuclide will come to the ground, so ground radiation level always come up, go high when we get the rain, but after two hours it's all flushed. But still plant is sealed or covered. Then every time we have in the Fukushima area more radiation is there than last year. So this year's radiation level is higher than last year. It means plant is still emitting all radiation or nuclide.

MALE: Okay. We don't need to go to Fukushima. We can stay right here in downtown Tokyo and get all the radiation you want, courtesy of the People's Republic of China.

KARAKAWA: This is another program. This is [00:11:00] concrete came from China, and it was mines over [00:11:08] mountain. And in Japan it was we had a short of sand when we developed too much. Then we bought the sand from China and it was radiated. So this area, this is called kids play area, and the radiation level is a [00:11:28].

And for nuclear response our government did not—Government always keeps saying Fukushima is safe, so even though now it's not safe and radiation level is higher than last year, but government released the border so everybody is now entering Fukushima [00:11:55]. And last year we checked parents and the kids dose last October, and we already have a high dose last year and the kids' level of dose is 10 times higher than the adult because of maybe it's absorbed quickly. And it will damage their DNA.

So, we tried to fix that problem. We measure the P24, P51, and the [00:12:30] to check dose, and this is just new technology and we already have the blood testing system. And so everybody is now checking their body their self, but the government is still saying it's safe, so we still have a miscommunication problem.

And this is a city recovery. Government has a recovery budget, but it means strange. Recovery means we cannot build new city, we must recover or build. The [00:13:08], because as it was it's very strange. So all Mayors, the city Mayors are confusing now. So they want to build a new city, because lots of land is already damaged and also [00:13:24] and no ownership. But in the past we had a difficulty to make a road straight, because lots of property owners there, but it's gone now and we can make a straight road. But the government is asking, "Please make a road as it was." So it's very strange.

And also now some city Mayor is building a new city without any government budget. So they are asking a private company to build the city and give a market. So, some city will be successful.

And in terms of the international assistance, Japan got lots of assistance proposals from outside, especially from U.S. because U.S. and Japan have a U.S./Japan securities treaties so U.S. must help if Japan got some damage and Japan must help if U.S. got any damage. So, we had lots of proposals from other countries, but most of our proposals was rejected by Japanese government. And also nuclear facility is high security facility, then most of the foreign country people cannot enter or our government don't want to enter. But finally Japanese government accepted some international assistance later, but it was too late.

And this is U.S./Japan joint response. We have a [00:15:14], a joint task force system and center at Yokota Air Force Base, but the Japanese government and the Prime Minister did not ask anything, so everybody just [00:15:25] by. So, unfortunately, lots of people have passed away by mismanagement.

And also this is a security treaty, and it was made after World War Two in San Francisco. And 50 years already passed that the history, so we may need to update. And also everybody is happy to change the system more effectively we can coordinate for disaster response, not for war purpose. And this is a simple situation or level of stage in emergency. We have a prevention, preparedness, response, and recovery, and the mitigation, and the detect and the report. This is just a simple system. And this is logic of Japanese system, so everybody feels strange. This is a disaster response and prevention system in Japan. Any disaster are prevented and guaranteed safe by Japanese government, and this is Japanese government measure, governmental role. So then citizens don't need to prepare, because if any disaster happens to citizen it's government's fault, so government don't need to escape or evacuate anything.

So, government also has an evacuation shelter, but it's not armed or nothing there. So if we go to our shelter we would be killed because no food, no service, no water. So, in last year disaster event happened and lots of people was killed by this system, so Japanese government is now changing. Rather than prevention to response and government don't need to guarantee any like a protection.

And this is another politics, and Japanese government are not honest to citizens to give information, because they're scared about facts and they want to control our information. And definitely our Prime Minister knew all the information, but he did not make a decision and also did not give information. And there is still we have a risk communication problem and also human rights problem. And the politics are in many place, and a big problem of Japanese politics and maybe our Prime Minister or our current other administration was wrong last year, but it's not only his fault. Japanese law, our constitution was wrong.

After World War Two the constitution emergency function was cut so we Japan cannot declare any emergency by law. So by law means with power and legal power and also budget we cannot do, because Japan was kind of an extension of United States after World War Two. Then the

U.S. cut this emergency declaration function from Japan constitution, so we are now trying to amend this declaration system. Maybe take the two, three years to change.

So, where Japan should go or what Japan should do is Japan constitution, emergency declaration amendment with U.S./Japan security treatment update. This is a plan and a disaster response concept transformation. Japan is maybe number one for disaster prevention, but people, humans cannot prevent all natural hazards, so we should think about we need to respond always. But as I explained, Japanese government said government is perfect so citizens don't need to escape, because government protection system is perfect, but with nature perfect is not possible. So we are changing prevention to response capabilities built. It's too late for Tokyo City. Tokyo City has now a high risk for new earthquakes. It will be maybe a magnitude of 9 or magnitude of 9.5 is expected, and near 100% risk in two years. And after two years past we'll have a 75% risk for 30 years, because Tokyo City are too much population, and also we have three layers of fault line, fault line plate, so it's a very dangerous area.

And also we made research capability of Tokyo Metropolitan City and medical service is not enough, and it's too late to prepare the medical services, so Tokyo City is now reducing population too much, but it's too late. But as much as possible we do it. So most of [00:21:14] is not in Japan, because if non-Japanese speaking people are in Tokyo we would have a risk and difficult to control, so we are limit residence people.

And also political leadership development is needed. So U.S. and Japan is now developing a joint leadership creation. It's too late, but we start and maybe 10 years or 20 years later we will have a more good leadership.

And this is the Japan nature in emergency system in the past. It worked before World War Two, but now this system does not work. Japanese are at risk always like earthquakes, tsunami, typhoon, and one event happened and caused a disaster, and we create wisdom and wisdom will make resiliency. Then Japan will— This is a nice circle and Japan was a resilient city in the past, but we have too much population so we are not good resilient system now, so we try to [00:22:33].

This is a simple [00:22:37] model. [00:22:39] and IMS and ICS. Those are kind of our people. And we can [00:22:47] good system.

And this is expected current damage in the expected Tokyo earthquake, Kobe earthquake and the 3.11 earthquake last year is completely different. And Kobe was a vertical shake and last year we had a horizontal shake, so post economy was 200 billion U.S. dollar both, and asset damage was 100 billion and 200 billion. And East Japan is wider than the Kobe city, so we have a three, four state in East Japan area. And when we get Tokyo earthquake we have a post economy is two trillion U.S. dollar economy in Tokyo, and this is a vertical, and the death will be 10,000 and in two weeks will have—We just calculated or simulated, and one million people will die in two weeks because of our medical services not functioning. So this is the expected earthquake and the new damage to Japan, so we try to set up quickly the emergency declaration function.

Without the emergency declaration—In Kobe earthquake we discussed emergency declaration is necessary, but everybody forget and we got the earthquake last year, so we got another similar damage. So for Tokyo earthquake [00:24:36] we try to do and change our constitution, change or amendment. And Fukushima is still active or in a disaster, and the Fukushima plant is 150 years program, and other areas 30 years recovery program. So just for the contamination it takes 150 years in Fukushima, but the people are still living there.

In a disaster situation we have lots of missing function and I have video and you can see some missing function in the disaster response. Just a moment.

[Video Plays]

KARAKAWA: This was originally a food and supply system. Lots of U.S. companies, like Big Lots and lots of companies are donating the product that we [00:27:20] back to this station with our U.S. [00:27:23] cargo and also [00:27:24]. And you can see this is a [00:27:31] disaster product, and the shower is also not included. So we found lots of missing product or services, we can say services. Also women need cosmetics, and if they don't use cosmetics they will get distressed, so we deliver the cosmetics also. And for kids we got a big donation from toy companies, so we delivered all toys to kids, and also school textbooks. And those products are not in a stockpile. Stockpile is just for water and food. Even power generator was not in the stockpile.

And this team will work for another 20 years. This is a plan. Most of our [00:28:42] people are assisting people or contribute, just go to damaged area once and they never come back and just visit one day, but this team is continuing to work for 20 years. And the government did not support them. We have funding by private sector people. Also survivors are happy to pay like for a haircut. They are happy to pay. Japanese people need massage. And we are also making a new business [00:29:21]. It's not for business purpose. It's everybody is happy to pay and everybody is happy to do a service.

[Video Ends]

KARAKAWA: So, we say this is a more political or social [00:29:38] when certain peoples [00:29:41] rather than one person one thousand feet. This is a concept of this team. So, this is a situation now in Japan, and lots of bureaucratic problem is there, and we try to fix, but it's very difficult to fix. Maybe next week we have a new Prime Minister. We don't know the same guy or a different guy, but we'll get a new Prime Minister and change old policies next week, and we hope they cover the disaster response, but we are not sure yet. There is the election. Thank you very much.

CHUCK: Thank you. Question time. We have time for just one or two questions. Tom, did you have a question?

TOM POPEK: Well, I do actually. So, again, this is Tom Popek. My question, here in the United States we have a public comment process for proposed regulations; in Japan is there a public comment process? And if there isn't is that something that your organization is encouraging?

KARAKAWA: Yeah, we have a public comment process, but most of the case it does not work; just we speak out, but our citizen people are just listening, but the citizens also don't have much capability to change policy. And when we try to change any policy it's very difficult. It's politics area. And so the reason why we initiated Japan [00:31:26] initiative, government does not function, then we asked our former Defense Minister on call and they belong to Japan Resiliency initiative, and we used retired or [00:31:43] soldiers. It's they are not active, but they are happy to do for social purpose.

CHUCK: Thank you. That's very similar to the InfraGard model we're producing by getting leaders of industry and government together to work on these issues. We have another question here.

PAULA GORDON: My name is Paula Gordon. Thank you for your presentation. And I also want to comment on Mr. Popek's—Is that right? I have been very deeply concerned since working at FEMA in the early 1980s and at National Science Foundation about some little known research that's been done into the vulnerability of particularly U.S. nuclear facilities which have been built according to structural engineering criteria, and not taken into adequate consideration mechanical engineering criteria. Therefore the magnitude of earthquake that nuclear power plants in the United States have been built to withstand is too low. It's much too low. It's been configured too low, because what can happen in an earthquake, according to the mechanical engineering research that I report on in an article that will become available in the next few days through the journal of the [00:33:19] security, what has not been taken into consideration is what can happen to rotor baring systems in a high magnitude earthquake where they can become centripetal forces and gyroscopic forces can become such that they will become projectiles and have a ruinous effect on nuclear power plant facilities.

Now this is material that has not been well, that people have not become aware of that was originally done by the National Science Foundation way back in the 1980s. What I would like to ask you is in your discussions with people in this country concerning the implications of the Japan earthquake and tsunami for nuclear power plant safety in this country, what have been your feelings concerning the level of awareness and the level of concern? Do you think it's anywhere near what it should be?

CHUCK: I guess you both can take a crack at that or either of you.

KARAKAWA: Basically for any hazard or response we should focus to response in the citizen safety, but in Japan's case nuclear is national security so they are more focused to how to protect the plant and rather than the saving the people, and that is another problem.

MALE: And I can respond somewhat to your observations about safety standards which are structural in nature but don't necessarily apply to equipment which then could produce cascading damage within a plant. So, there is a similar situation with U.S. nuclear plants in that they have what are called GSU transformers, Generator Step-Up transformers. And should they be subjected to geomagnetically induced current from a solar storm they can overheat and in some cases explode and catch fire, which is a safety concern. So I think your general point is just looking at these safety issues from one dimension, for example a structural dimension, really doesn't encompass the whole range of threats.

And I would also point out that again the vendors for the Japanese plants at Fukushima were essentially U.S. vendors, and so we all have a similar problem still.

CHUCK: Thank you very much. And we're running a little bit over time. Let's give our guest from Japan a round of applause. And hopefully they can stay as late as the last hour where we'll have some more opportunity to have some give and take.

An Overview of the Vulnerability of Electric Power Grids to EMP, Geomagnetic Storms, and Non-Nuclear EMP Events
John G. Kappenman, Storm Analysis Consultants

Introduction
The High-Altitude Electromagnetic Pulse, Severe Geomagnetic Storms, and Non-Nuclear electromagnetic pulse (EMP) (or Intentional Electromagnetic Interference) fall into a category of High Impact, Low Frequency (HILF) events: risks whose likelihood of occurrence at any one time are low relative to other threats, but which could significantly impact modern electric technology systems, in particular high-voltage power grids, were they to occur.

Overview of EMP and Geomagnetic Storm Risks
The EMP risk comes from a variety of possible scenarios involving the detonation of a nuclear weapon at high altitudes (above 30 km). This could be the result of an intentional attack carried out by a rogue nation or a terrorist group. It is conceivable it could also occur as a result of the interception of a nuclear missile at high altitudes.

Recent analysis carried out for the EMP Commission, FEMA, FERC and the U.S. National Academy of Sciences has determined that severe geomagnetic storms (i.e., space weather caused by solar activity) has the potential to cause crippling and long-duration damage to the North American electric power grid or any exposed power grid throughout the world. The primary damage impact to the power grid is the risk of permanent damage to high-voltage transformers, which are key, scarce, and difficult to replace assets for the high-voltage power network. Other important are also potentially at-risk, but these risks are not as well understood. This would be an event larger and more damaging than hurricanes with financial impacts exceeding $1 trillion per year and with recovery times that could span 4–10 years, which would place many lives at-risk.

These storm events can have a continental and even planetary footprint causing widespread disruption, loss, and damage to the electric power supply for the United States or other similarly developed countries around the world. It is also estimated to be plausible on a 1 in 30 to 1 in 100 year time frame. In short, this is potentially the largest and most plausible natural disaster that the U.S. could face, as the loss of electricity for extended durations would mean the collapse of nearly all other critical infrastructures, causing wide-scale loss of potable water, loss of perishable foods and medications, and many other disruptions to vital services necessary to sustain a nation's population. It is important to note that this naturally occurring phenomenon attacks the electric power grid in a way similar to that posed by the late-time (E3) portion of an HEMP attack. In addition, missile defense systems would not be effective against the source of this phenomenon, which is the Sun. Therefore hardening the electric power grid is needed and also provides effective protection against both the E3 HEMP effect as well as against a severe geomagnetic storm—a two for the price of one solution.

Other valid scenarios with varying impacts include high-power non-nuclear electromagnetic weapons (EM weapons), which have a limited area of impact unless used in a coordinated attack. These weapons will produce intentional electromagnetic interference (IEMI) or Non-Nuclear EMP. These events have the potential to physically damage electrical and electronic equipment and/or components, most notably industrial control systems like SCADA. This attack scenario is considered to be nearly as likely a threat to the electric power grid as that posed by cyber attacks on this infrastructure. Also this electromagnetic threat is similar in nature to that posed by early-time portion (E1) of an HEMP attack. However, it also should be noted that missile defense systems would not be effective in preventing this mode of attack, and that the only preventative measure would be the hardening of the exposed elements of the power grid. Therefore there are again "two for the price of one" benefits, in that hardening the power grid to withstand an EM Weapon will also provide similar protection of these critical infrastructures to E1 HEMP. In all cases, the most significant concern is the potential for a simultaneous impact to large portions of the electric power system, from which restoration and recovery may be challenging and prolonged.

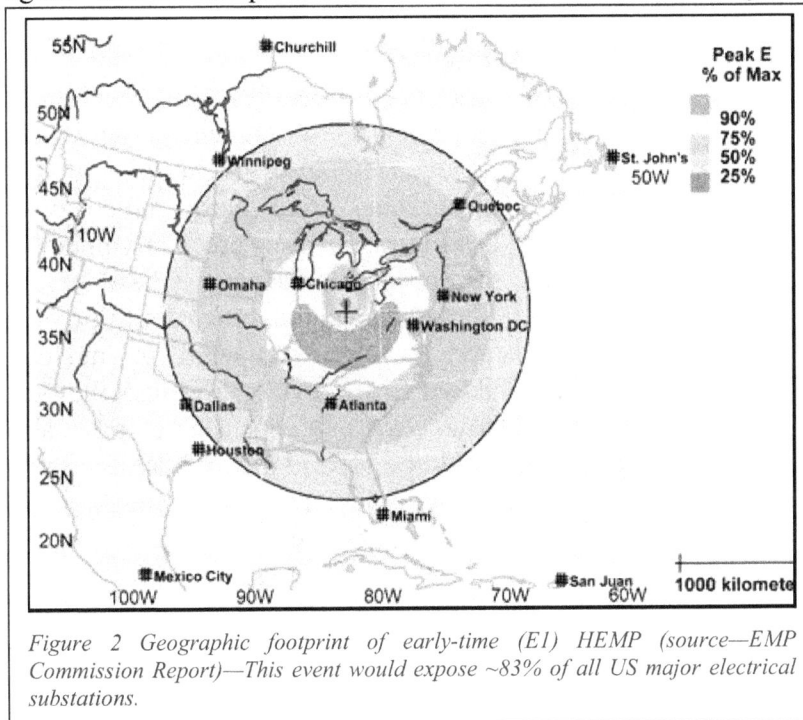

Figure 2 Geographic footprint of early-time (E1) HEMP (source—EMP Commission Report)—This event would expose ~83% of all US major electrical substations.

The severity of the threat from EMP, Non-Nuclear EMP, and Geomagnetic Storm impacts to present day electric power grid infrastructures around the world have grown as the size of grids themselves have expanded by nearly a factor of 10 over the past 50 years, while at the same time they have become much more sensitive as microelectronic control systems have supplanted prior generations of electromechanical relays and vacuum tube components which could withstand higher levels of EM radiation. These aspects of current design practices of electric grids have unknowingly and greatly escalated the risks and potential impacts from these threat environments. There has been no power grid design code that has ever taken into consideration these threat concerns, yet it is possible to remedially apply relatively inexpensive solutions and protocols to harden both the current power grid and to also add hardening to future additions to this critical infrastructure.

EMP Risks to the U.S. Electric Grid

In a series of reports that I had authored for the U.S. FERC, we had developed detailed simulations and assessments of the vulnerability of the U.S. electric power grid to EMP attack threats. Figure 3 provides a map derived from Figure 2 of the E1 threat footprint of the locations of all HV and EHV substations that would be within direct line of sight exposure to the

E1 pulse. As shown, the laydown of the E1 environment would cover ~83% of all HV and EHV substations within the United States. In addition there are estimated to be another ~40,000 distribution substations with similar ratio within this footprint.

Figure 4 provides a similar assessment only for the electric power generation plants across the United States. In this case ~75% of all U.S. electric generation facilities would be exposed to the E1 EMP pulse. Figures 2–4 demonstrate the potential that exists for substantial impacts to the U.S. power grid. Although further detailed simulation and analysis models not described here provides further confirmation of the potential for widespread catastrophic damage to various microelectronic based equipment in modern power grid infrastructures. These systems have not been specified or specifically engineered to contend with these unique impulses. On a smaller geographic scale these same power grid infrastructures are also vulnerable to non-nuclear devices. These have a shorter range, but are easier to procure and develop compared to nuclear-based EMP devices. Because they are low cost and concealable, they also could be used in a coordinated fashion at multiple locations to attack major portions of the electric grid.

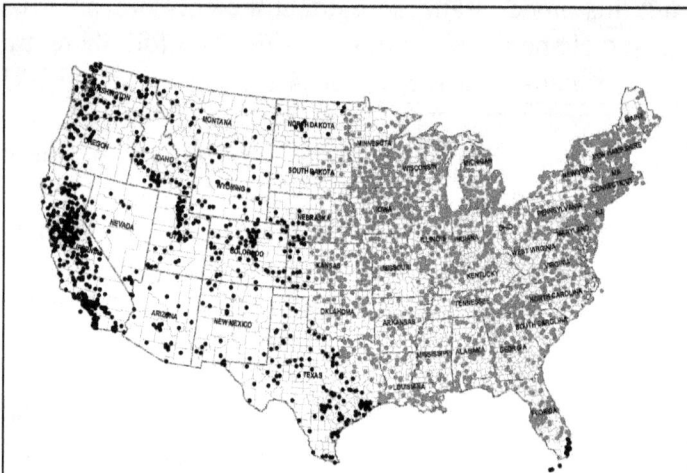

Figure 4 HEMP Fast Pulse exposed power plants (Red) total 10,730 with a generation capacity that is ~74.4% of the U.S. total generation capability.

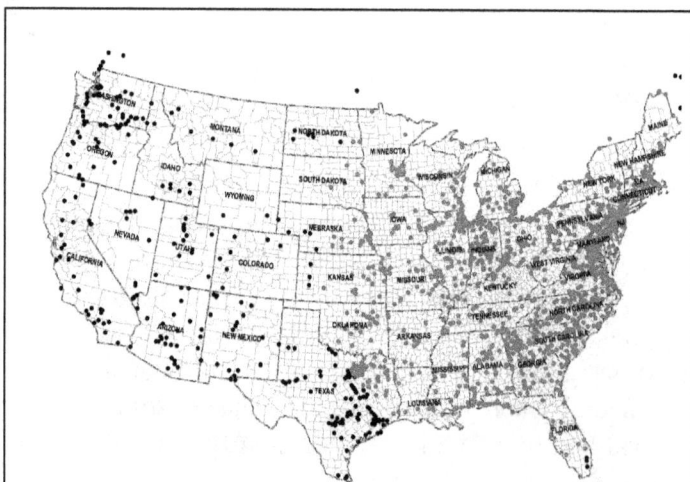

Figure 3 - HEMP Fast Pulse Exposure covers a total of 1765 substations exposed or ~83% of 2106 major HV and EHV substations.

Similar levels of damage potential exists in the U.S. electric grid from large geomagnetically induced current (GIC) flows caused by either severe geomagnetic storms or the similar type of wide spread geomagnetic field disturbances associated with the slow pulse or E3 portion of the EMP. Both the E3-emp environment and severe geomagnetic storms can cause large geomagnetically induced currents (GICs) that may damage power grid equipment, such as transformers, generators, circuit breakers, and capacitor banks. While GIC from geomagnetic storms are known to cause damage, the E3 threat environment will cause very large short duration GIC as well. Basic vulnerabilities to both E3-caused and natural geomagnetic storm disturbances have been growing largely unchecked as the power grid infrastructure has grown in size. Damage to the power grid infrastructure can cause the possibility of an extremely slow pace of restoration and the multiplying effects that could cripple other infrastructures such as water, transportation, and communications due to the prolonged loss of the electric power grid supply.

Mitigation Technology Options of the Electric Grid
In both protecting against GMD and EMP there have been a variety of new technologies that are emerging to provide new and more effective means of reducing exposure to EMP threat environments and the resulting vulnerability of this critical power grid infrastructure.

Transformer damage is the most serious concern that could develop due to E3 HEMP threat scenarios or severe geomagnetic storms, although other key assets on the grid are also at risk. In particular, transformers experience excessive levels of internal heating brought on by stray flux and circulating currents when GICs cause the transformer's magnetic core to saturate and to spill flux outside the normal core steel magnetic circuit. Previous well-documented cases have noted heating failures that caused melting and burn-through of large-amperage copper windings and leads in these transformers. These multi-ton apparatus generally cannot be repaired in the field, and if damaged in this manner, they need to be replaced with new units, which have manufacture lead times of 12 months or more in the world market. In addition, each transformer design (even from the same manufacturer) can contain numerous subtle design variations. These variations complicate the calculation of how and at what density the stray flux can impinge on internal structures in the transformer. Therefore the ability to assess existing transformer vulnerability or even to design new transformers to be tolerant of saturated operation is not readily achievable, except in extensive case-by-case investigations.

Current estimates have been made that over 300 large power transformers in the U.S. grid alone could be at risk. This includes approximately one-third of the entire fleet of 500 and 765 kV transformers currently operating in the U.S. grid. The 500 and 765 kV transformers are the backbone of the grid that extends into regions that contain nearly 80% of the U.S. population. As discussed in recent FERC reports, several options were reviewed to counter potential damage and delayed restoration brought on by the E3 and severe geomagnetic storm environment scenarios. These can be summarized as:

- provisions for replacement equipment (i.e., spare transformers, circuit breakers, etc.);
- blocking devices in the network to block and completely prevent the flow of GIC (primarily transformer neutral to ground blocking devices, but also transmission line series capacitors); and
- low-ohmic transformer neutral-to-ground resistors to partially reduce GIC levels.

Conceptually it is possible to use capacitors installed in the neutral to ground lead of transformers to block the flow of all GIC from entering and exiting an electric power grid and its transformers. Such devices in actual application are not simple and would require highly reliable and rapid bypassing of the capacitor under normal AC system fault conditions to prevent harmful overvoltages that could damage the device itself as well as the transformer. These device concepts were successfully developed and demonstrated under an EPRI research program of which John Kappenman was the PI back in the early 1990s.

The Kappenman Method of GIC blocking device for application on bulk power systems requires both a capability to block DC or GIC flows into or out of the transformer through utilization of the resistor while simultaneously allowing a path for the flow of long-duration low-level AC currents and very short-duration but very high magnitude AC currents due to nearby faults on the high-voltage network. Figure 5 provides a more detailed schematic as it would be installed in a typical transformer neutral to ground connection. As outlined in the schematic, the device entails two primary circuit systems, a capacitive circuit (or as an alternate a resistive circuit) that provides the GIC blocking and normal flow path for low-level long-duration AC currents (such as due to phase unbalance). The other circuit system is the bypass circuit that provides the rapid and temporary bypass of the blocking resistor under AC fault conditions. This bypass circuit provides the path to ground for the short-duration but very high magnitude AC currents that will be created during a nearby AC fault conditions.

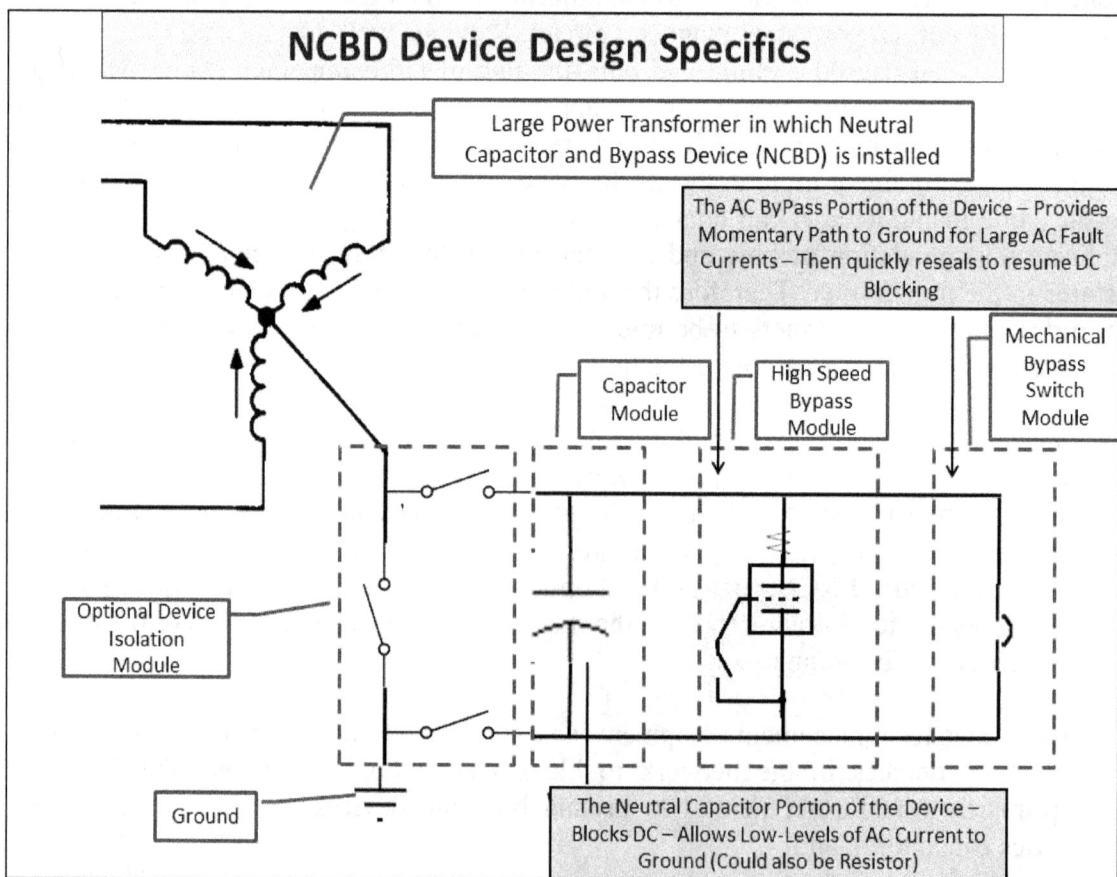

Figure 5. *Schematic of a Kappenman Method Neutral Capacitor and Bypass Device (NCBD) in a transformer neutral.*

While this approach reduces the power grid exposure and vulnerability to GIC caused by either severe geomagnetic storms or the E3 portion of the EMP pulse, other separate hardening approaches are necessary to protect key grid electronic systems from the E1 portion of the EMP pulse or fast pulses from non-nuclear sources. In this case a combination of already known technologies (Faraday shielding) and new technologies need to be deployed. Several highly promising approaches can be described in more detail here. Recently, though, an old technology has been revived that appears to have all of the right qualities: the trusty vacuum tube. A start-up called Advanced Fusion Systems (AFS), based in New Rochelle, NY, has developed a family of protection devices that uses a high-power vacuum tube. Not mentioned in discussion of the GIC blocking device is a key component for fast-bypass which is an AFS electron tube embedded into the overall schematic shown in Figure 4. This same technology can also be deployed for other system protection purposes for the Fast transient. This protection capability is enabled due to the inherently fast turn-on operating speed in the 10's of picoseconds, fast enough to shunt the transient voltage rise to prevent it reaching a sensitive microelectronic device. These operating times are considerably faster than previously available MOV technology. In addition electron tubes can absorb several orders of magnitude higher surge energy than MOVs as well, this combination makes them ideally suited for this important task.

Any transformer whose BIL rating is less than the applied field strength from and E1 impulse is vulnerable. These transients will also have steep fronts with rise times in the nanosecond range; therefore, the ability of the electron tube to operate rapidly is an important aspect in assuring protection. These devices are suitable for AC and DC circuits of any voltage within the stated range. Bulkhead mount configurations are available to protect shielded facilities for power feed and control signal feed through requirements as well. These are rated to meet all transient suppression requirements. Single devices will handle the full current load for incoming power supply needed for the control house facility in virtually all applications, as the ability to produce these devices in all distribution voltages and currents in the range of hundreds of KiloAmps are easily accommodated. Figures 6A and B show the patent-pending devices for bulkhead feedthrough (Figure 6A) and field collapse operation (Figure 6B).

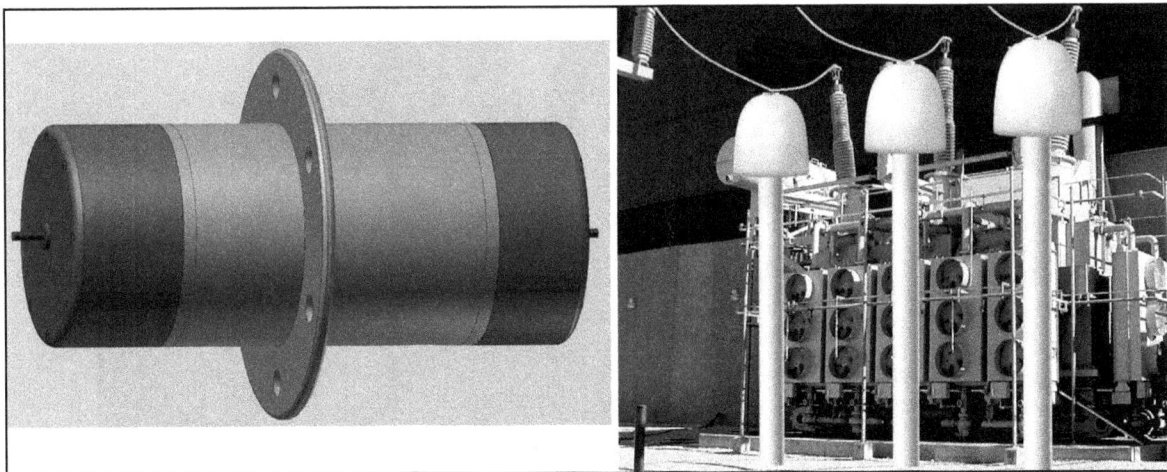

Figure 6(A) Bulkhead-mounted 4138 Bi-tron™. (B) Field-Collapse™ EPS™ (3-Phase Artist Concept).

While bulkhead protection devices can fill an important need on exposed power feed and other signals that need to be brought into a control house, it is even better to eliminate these signal wire penetrations if at all possible in the first place. Here again is a new solution that is now being developed for the electric power industry that takes this exact approach. Inside all major substations it is necessary to monitor the voltage potential and current flows on the HV and EHV power grid lines and transformers, so that proper sensing and control actions can be performed. In most cases this sensing is done via conventional transformers and metallic data wires which can readily couple with the E1 pulse environment and act in a way to collect and deliver harmful

E1-caused transient overvoltages to sensitive control equipment within the control house. The innovative approach that SmartSensCom has taken is the base this entire transformation and data communication technology on Fiber Optic conduits rather than metallic wire signaling. Their Electrical Phenomena Cluster (EPC) completely eliminates the possibility of the E1 transient being conducted into the control house and control house equipment in the first place. This sophisticated fiber optic-based technology is adapted from U.S. Naval developed technologies that have a long pedigree and have been employed in extremely harsh environments including undersea communication cables. Figure 7 provides a photo of the prototype device that is being tested in an electric utility substation. In addition to monitoring all AV voltage and current quantities with this device, it also has the ability to directly monitor the near-DC frequency GIC flows at HV and EHV voltage levels which is exceedingly difficult to do otherwise and provides important real-time situational and forensic information to electric utilities' of this dimension of the threat. The EPC approach also is estimated to be lower life cycle cost than conventional technology.

Figure 7. Electrical Phenomena Cluster (EPC), which can optically measure current, voltage, and temperature. No active electronics for E1 vulnerability instead optical cable transmits light signal to and from the control house. No conductive cabling required.

Conclusion

Today's electric power grids have been developed over many decades. However this development has brought with it enormous growth in vulnerability to both natural and intentional electromagnetic threats posed by EMP and GMD. These threats have grown to a dimension where they have been described as an "Unrecognized Systemic Risk." In part these risks have migrated into electric power grid infrastructures due to the fact that their development was not governed by any rational design code that took these threat environments into consideration. The U.S. FERC has recently issued a NOPR for the GMD threat that would require deployment of hardening measures to the present day grid. This paper provides a summary and overview of technologies not only capable of reducing electric grid GMD vulnerability but also addressing EMP vulnerabilities. Further since they offer a good combination of low costs and improved performance they make sense for application in all future power grid development.

Local Preparation for Long-term Nationwide Disasters

Mary Lasky, Chairman, Howard County Community Emergency Response Network (CERN)

Thank you for asking me to speak at this 2012 Dupont Summit. The main points I want to discuss today are:

1. Emergency and no-notice situations are happening more and more frequently now in the United States. In the DC, Maryland, and Northern Virginia area we had both the Derecho in June and Hurricane Sandy in October this year; Sandy gave us notice but the Derecho did not. There are shootings happening across our country (as I am writing the remarks of December 7, we have all be horrified by the mass shooting at the Sandy Hook Elementary School in Newtown, Conn.) The potential for terrorist activities increases as the United States seems to be attracting enemies.
2. Federal, state, and local governments may not be able to help quickly. The first responders may need to deal with their own families, roads may not be open, power may be out, etc. The officials will definitely have to prioritize and take care of the most needy first.
3. Citizens and businesses need to be prepared.

During Hurricane Sandy, New York and New Jersey did have several days of advanced warning that the storm was coming, but they were not prepared for the massive devastation, fires, loss of electrical power, and flooding that happened. There were millions of people and businesses without electricity and some of the hospitals in that area still are in only limited operations six weeks afterward. Some individuals are still suffering and businesses are struggling. Because other states were not hit so hard, they could respond—help was forthcoming. We may all know someone who responded and went to New York or New Jersey to help.

What if it were a wider area or the whole county? Then there would not be people to respond and come to the aid of the people and businesses. With something so massive, government may be overwhelmed and not be able to respond quickly. For example, if there were a massive Coronal Mass Ejection from the sun or an EMP or a coordinated cyber-attack, then all our power would be affected—government, businesses, and individuals. Are we prepared as a government to handle this? Are businesses prepared? Are we as individuals prepared at home to sustain ourselves for days, weeks, or even months?

Our responsibility as citizens and businesses is to be prepared. For businesses—this means have a plan of what to do during such an event. This is called continuity planning. There is a national organization whose members are trained and certified to help with this type of planning—it is the Association of Continuity Planners (APL). At APL, we have a 100-page pandemic response plan, we have continuity plans for every department, and a 100-page crisis management plan. Yet, even with such detailed plans, *we have found a high level, one-page "trigger" document to be the best way to get your mind focused and then directs us to specifics in the big plans. The one-page document gives us a good feel of what happens at different stages of the event.*

The Community Emergency Response Network (CERN) in Howard County, Maryland had a committee work on preparing a way for businesses to be prepared. Because the majority of the businesses are small in Howard County, we wanted an approach that was efficient and effective and did not take long to accomplish. The results of this work are available to everyone. On the CERN website, there is a one-page document that helps an organization work through its continuity planning. It helps guide a business to look at risks and how to mitigate them, looking at information technology and staffing. There is a second one-page document, which is a sample plan that businesses can use as a template for their own plan; for example, putting in where they would go if they had to establish an alternate location. The back of this form can be used for important telephone numbers of your staff, banks, insurance companies, etc. I have worked with nonprofit organizations in their planning and they have been able to create a workable plan in no more than two hours. Any businesses can easily devote that amount of time to be prepared. The CERN website is located at: www.cernhc.org.

As citizens, we need to be self-sufficient. So what should or could we be doing? We need to be prepared for situations where help will not come for a few days or a week or a month. We need to have supplies at home that will sustain us for at least two weeks, a month, a year maybe. The Church of Latter Day Saints has a tenet that the members of the church need to have enough food to last a year; in fact, ideally a two-year supply so they can help others. One of the other important things we all need is a communication plan with our family. In addition, we may need to evacuate our homes quickly and having a "Go Bag" may save your life. Good lists can be found at www.Ready.gov. Red Cross also has a good list, and CERN has good lists on its website, too. If a massive EMP leaves us without with no power for eons, it would be ideal if individuals could generate some of their own power and grow some of their own food. The idea is: Help yourself be sufficient and then be able to help your neighbor.

Again, emergencies with and without notice are happening and will continue to happen. Federal, state, and local governments may not be able to be there to help quickly and citizens and businesses need to be prepared and take care of themselves.

DoD HILF Role
Paul Stockton

Assistant Secretary of Defense Dr. Paul Stockton *is responsible for the Department of Defence (DoD)'s support of civil government and related homeland defense support. He walked through the concern that DoD has for these issues not only for military bases and the ability of DoD to conduct its mission but, for the country as a whole that is equally dependent of civilian critical infrastructure for its operations. He emphasized the need for military bases and local communities to become more sustainable at the local level and shared experiences of failing to do enough in these areas. He profusely thanked InfraGard for keeping these conversations active between the private sector, federal, state, and local authorities.*

CHUCK: Thank you, Jeff. Right before we introduce folks I'd like to ask a question. How many people of a show of hands have read the National Preparedness Report of this past March, this past March National Preparedness Report? Okay, I see about seven hands. It illustrates something very interesting. They have a paragraph on the bottom of page five. If you read nothing else I would encourage you to read it. It talks about their worst-case scenario. They don't tell you much about it, but they tell you two things, and I'd like you to think about it as you hear the next speakers.

First, every local community—and they define local community as an area with seven million people—(so draw a circle anywhere in the country that holds seven million people) should be ready to handle 265,000 medical casualties. Now if you think about any of the hospitals you know you will not find an area of the country that would have more than 20,000 hospital beds. What do you do with the other 240,000 people? And the second thing this report says, and you must be prepared to do this with minimal assistance from the government. Why might that be? They don't tell you why that might be.

Now I know if I woke up in the morning and my city of Chicago was on the news because somebody had detonated a 10–20 kiloton weapon off the back of a pickup truck and a quarter million people died and a quarter million people got sick right away, as horrified as we would be half the country would be coming running to their aid. We would have to keep people out, because we love each other that much and that's how we take care of each other in this country.

So why isn't somebody able to come take care of them? And I don't know because the report doesn't tell me. I can only guess one thing. They are being very courageous and daring and introducing in a very quiet way some horrendous scenarios that politically are impossible to talk about, like these, because there may be scenarios like that where all the caring, capable, wonderful people who want to come rushing to your aid just can't so you have to be ready to take care of an extra 240,000 medical casualties on your own. And yet we haven't planned that and we haven't exercised it and we haven't trained for that, at least not to my knowledge.

Now to the great credit of the National Defense University (NDU) and the Defense Department and a few others a little over a year ago for the first time a comprehensive training workshop and exercise was done at NDU with all of these folks involved, and a broad range of federal, state, local government agencies as well as the private sector. And it was the first time we looked at anything that might be a collapse of infrastructure nationwide that would last more than a month.

And so what I would like to do now at this point is introduce the person who is going to introduce our next speaker, Dr. Lin Wells. He is a fellow at National Defense University (NDU). Most of you would never know him, but I can tell you that I've never met a guy who would be brighter, broader in scope of understanding of issues in life, and more caring than he. And with that, I thought he would be the most appropriate person I know to introduce our next speaker. Dr. Wells.

DR. LINTON WELLS: Chuck, thanks. Let me be very brief. Dr. Paul Stockton, Assistant Secretary of Defense for Homeland Defense and American Security Affairs was confirmed by the Senate in 2009. As the Assistant Secretary he is responsible for supervising DOD Homeland Defense activities, which includes defense critical infrastructure protection and mission assurance. The word mission assurance is very important activities. He does defense support of civil authorities, defense crisis management in Western Hemisphere, security affairs. The rest of his bio you can read in the conference materials. Without further ado, let me introduce Secretary Stockton. It's a pleasure, sir.

PAUL STOCKTON: Thanks, Lin. Thanks for everything that you do at NDU to help lead the charge. Chuck, I am enormously grateful to you and InfraGard for the opportunity to speak today, but more important for the leadership that you're exercising in this absolutely vital realm. So thanks to you, thanks to everybody in the room to make sure that this issue gets the prominence that it needs for policymakers, but also all of you building solution sets that are going to be viable, that are going to politically make sense. It's absolutely critical that we do that.

And in that regard, I want to apologize, Jeff. Where are you? I cut into your time. That BENS study is extremely important. You know what I got out of it most and that is the need to think deeper about the finance side. It's not only a technical side. It's from a business case model how do we get the financing that it is going to be needed going forward? So thanks so much to you and all of your colleagues for that.

I want to talk a little bit about how my portfolio has changed, how it applied to the problems of infrastructure in general and then electromagnetic pulse (EMP) preparedness, solar storm, geomagnetic disturbance preparedness, all of those threat vectors that we are together concerned with. I want to talk about first response and then preparedness. I know that's a little backwards, but that's the way my brain works, and it works this way because we are in the midst now of our after action review from Hurricane Sandy.

Folks, we suffered a strategic surprise in Hurricane Sandy. We did not understand, industry did not understand the degree of fragility and especially interconnectivity between the power grid, the electric power grid, and the gas distribution system from the colonial pipeline to the buckeye pipeline, everything else. We were in discovery mode in the midst of a disaster. We had to build the airplane while flying it. What I mean there is that we had to build a concept of operations of how the Defense Logistics Agency was going to provide absolutely vital support for emergency power in order to save and sustain lives.

So, when you look at what that category of activity included we are doing many things for the very first time. We had giant transport aircraft, C5A aircraft for the very first time in DOD industry loading up utility trucks, cherry pickers in Southern California, in the state of Washington, in Phoenix and flying them to the East Coast so Con Ed and other industry partners could make use of them for grid restoration. We had the Army Corps of Engineers, absolutely terrific folks working together with their contractors. We had the Defense Logistics Agency supporting this effort to get emergency generators installed where most needed.

And this comes as no surprise to you, but it surprised a lot of folks up in the disaster area that if you run your emergency power generator for more than 48 hours guess what, they're not engineered for that, they're not maintained for that, they are going to break down. And so if we look at the kind of long-term disruption of the power grid that we could experience due to the threat vectors that form the basis for this conference we need to think about what it is going to take to bring capabilities in as part of the disaster response, what kinds of capabilities are going to be needed, large-scale emergency generators, power installation, and then above all the flow of fuel.

We were providing fuel to six critical facilities for Verizon so they could maintain cell phone coverage in the New York metro area, because guess what, they didn't have the plans in place to get the backup fuel they needed once their commercial providers could no longer execute that critical function.

The interconnectivity of different components of the energy infrastructure, gas and electric, the cascading failure of critical infrastructure, starting with electricity and then moving down into communications and transportation—Folks, I'm going to ask all of you to think about what the effects will be above all for life saving and life sustaining if due to EMP or geomagnetic disturbances the grid goes down hard and we're back, I'm afraid, into discovery mode.

I don't want that. I want to have thought about these factors, these common points of failure now today with your leadership to understand how, from my defense support to civil authorities perspective, how I can support Department of Energy (DOE), how I can support FEMA and Department of Homeland Security (DHS) so they in turn can help the nation's Governors and Mayors save and sustain lives. We're in a big after action review process here inside the Department of Defense (DoD). Later we'll want to turn to our inner agency partners and ultimately, of course, the perspectives of industry are going to be absolutely vital.

So for the very first time I'm asking questions —shame on me for not having asked them before. Given the kinds of emergency management compacts and agreements that exist between utilities that they use against traditional threat vectors where they come in and support each other, sometimes from considerable distances, okay, how can the Department of Defense (DoD) and our federal partners be prepared to support that for the kinds of events that you are discussing today which could produce very, very long-term outages? What do we need to help support industry in a way that makes sense for the limited role of the federal government in these kinds of events?

So there is a whole realm of disaster response that we're digging into. I think Sandy just gave us a taste both of the nature of the response requirements that could emerge and then the kinds of demand-pull that would come on the Department of Defense, my personal interest, my personal responsibility. But how would that again be worked out in advance with industry as well as our federal partners? This is the new realm. I'm calling it the new dimension of defense support to civil authorities. And I would ask you all to help me build out this new dimension, understand it, be prepared for it, especially in the context of EMP and geomagnetic disturbances.

Of course, cleaning up on aisle nine, doing the response, we would like to avoid that, we would like to do prevention and mitigation from the start so that the demand pull for this defense support, so that a threat to human lives in these kinds of events is lessened before they occur. And that's an area where I think we need to continue to press forward. I think, Jeff, some of the points that you made about continuing to strengthen policy guidance, build more coherence in what kinds of things we want from the Department of Defense perspective in terms of working together with our lead federal partners, DOE and DHS, thinking about how we want to build partnerships with utilities so that we can make sure that critical defense facilities are able, again my worm's eye view, execute the core DOD missions that the President is going to direct us to do even if the grid goes down, even if the grid is disrupted due to geomagnetic disturbances or EMP.

We need to be prepared for that. There are certain things the Department of Defense can do inside its own facilities, inside the fence line if you will, but much, much more needs to be done in partnership with industry, knowing that the Department of Defense depends for 99% of its electric power on the commercial sector. There is only so much we can do inside the fence line. Much of what we need to be doing is in partnership with all of you in industry, with always, always the Department of Energy (DOE) and the Department of Homeland Security (DHS) in the lead for these partner activities, because they are in the lead for the federal government over issues related to critical infrastructure that is owned by the private sector, and above all energy infrastructure.

So, we've got a lot of work to do. We've got our work cut out for us. On the other hand, again, there is progress being made. There are opportunities to build, for example, peaking power plants on DOD land inside our installations that ordinarily would sell their power back to neighboring communities. Utilities would make a good return on investment, but if through the kinds of threats you're discussing here at this conference the grid were to go down then that would be a source of power to execute our critical missions in those facilities. There are all kinds of opportunities to make progress in partnership with industry that reflects the fact that industry needs to have some way of recovering costs for investment in resilience against these nontraditional threat vectors. Let's get rolling, because again Sandy is proof that we need to accurate our preparedness.

And I'm as guilty of this as anybody else. In fact shortly before Sandy struck, just a few weeks before Sandy struck Congress has given us in the 2010 National Defense Authorization Act the authority for the first time to use our Title 10 reserves in responding to national disasters. A really important authority, we finally got it. So, during the course of this past hurricane season we are pushing forward to try to get at least some interim measures so we can mobilize the reserves, bring them into the fight. And then around October I said, "Okay, well we're pretty well through this hurricane season. I guess we can turn to some other priorities and we'll get this one, we'll keep it on a burner, but we'll kind of move it to the back burner, because hurricane season, it's winding down, isn't it." And then Sandy struck.

Shame on me, but I think it's a metaphor for the need to sustain progress, to keep fighting, keep pushing against the resistant bureaucracy that sometimes we encounter. And I want to thank all of you for helping me do that, because you can lead, you can do things that are very difficult for those of us on the inside to be able to pull that off. And Chuck, in that regard I want to thank you, all of your InfraGard partners for helping lead the charge. Leadership is what is needed. You're providing that. Chuck, thank you so much.

And with that, I have kept my remarks very brief so I can have your perceptions on what I should take back, your recommendations on what I should be doing within the Department of Defense in working with our lead partners, DOE and DHS, and above all thinking about how in my own portfolio of mission assurance, that is ensuring the Department of Defense can execute its core missions, even if the grid is threatened by these new vectors, we can ensure that we can do what the President wants. So with that I'll open it up for any questions, any recommendations. I welcome that very, very much.

CHUCK: I have a microphone. And come to me and I'll meet you halfway. Is Dr. Wells still here? He could have a first question since he introduced here, and then others can line up behind me.

DR. LINTON WELLS: Thank you, Mr. Secretary. Back in the 1990s with the Presidential Decision Directive 63 the idea of information shared in analysis centers and building this long-term partnership between government and industry has been very checkered success, and some success was done very well. What have we learned out of Sandy and some of the things that might allow us to do this better going forward in terms of the government/industry partnership?

STOCKTON: Well we have relearned a lesson. I guess if you don't learn the lesson it's not really a lesson learned. We have relearned the lesson that more sharing is going to be better and that industry itself may not have the information that is required for us to be able to provide effective defense support, so there needs to be information gathering and information sharing, with one significant exception. The gas companies in the New York and New Jersey metro area could not tell us which gas stations had product but no electricity where we could install generators to get that product pumped into vehicles or conversely which gas stations had plenty of power but no gasoline where we could truck the gasoline, fill it up, and again they would be good to go. We couldn't get that information out of industry, because industry didn't have it.

So, we not only need better information sharing, we need to gather the data, we need to understand which terminals for the distribution system, which connections in the gas distribution system need to be stood up in what order so that we can provide effective defense support. We need a better understanding of the data set and then we need to share that data, and not by making things up literally in the middle of the night as I was doing during Sandy, but have a regulized(sic) flow of information, have plans in place so that when and not if the next catastrophe strikes and there is an energy dimension to that catastrophe we'll be ready to go.

SHALOM FLANK: Good afternoon, Secretary Stockton. I'm Shalom Flank. I'm the Micro Grid Architect at Pareto Energy, a micro grid startup company here in D.C. You mentioned the importance of the financial and business side in order to be able to create reliable power infrastructure for defense installations, but there seems to be a disconnect in current DOD policy. There are new energy installations going in at a number of bases, for example a 15 megawatt solar installation that we had done some work now, where the energy is not usable by the base in the event of a grid outage. The necessary capability to make that energy usable takes a little extra money. The base has no budget authority to invest that increment in order to be able to use the capability that is being installed. What do you think is the best way to overcome that barrier?

STOCKTON: Well I think first of all making sure that we are aware of the problem, as you are doing right now, that's absolutely vital. I have partners in the Department of Defense, Sharon Burke, in operational energy, John Conger, in installations and environment, the Office of General Counsel, Bob Taylor. We're all working together very, very closely, really in an unprecedented way to make sure that we understand what are the emerging problems, share best practices, share bad lessons learned, understand where gaps are. So from a DOD-wide perspective we're able to bring this knowledge to bear and begin to make more progress than we have made in the past.

So, we are getting better aligned inside the Department of Defense, thanks especially to John Conger and Sharon Burke for making that possible. We're getting better and we're going to be a better partner for you. So thanks.

ROSEMARY ORSINI: Hi, I'm Rosemary Orsini. I'm a retired DOD employee and now I'm a concerned citizen trying to see how I can help my community prepare. And I appreciate all the planning that you as DOD organization is doing to perform better in the next event, but have you considered the fact that most vehicles produced since 1979 won't even function then? And how will you accommodate that issue for DOD and the nation?

STOCKTON: Well, that is an enormous challenge and it's one of many. We've got legacy infrastructure and components of infrastructure, POVs, trucks, etc. across the board that are not hardened against some of these threat factors. And so again let's get going. Let's think from a realistic way about what are the political impediments to progress and begin to take that on, knowing that there isn't going to be a huge amount of money that Congress is going to be able to appropriate against these problems in the current budget environment. That ain't going to happen.

So, in partnership with industry, in partnership with American citizens what kinds of progress can we be making, again with DOE, Department of Transportation, the other lead federal agencies helping to lead the charge with DOD in support? But there is a DOD internal component to that which we take very, very seriously. That is we need to be able to execute our missions from a continuity of operations perspective. We have got to be able to function in this environment, and we're paying plenty of attention to this and preparing against it. Thanks.

ANDREA BOLAND: Hello, I'm Andrea Boland, State Representative from Maine. We met in London. I just had a brief suggestion which would be to take your message on the road. I have really been doing a lot of work in the Maine legislature to try to bring this awareness of this issue forward and bring some legislation to deal with it as far as a state can. So, I really think there is an opportunity there to address even state legislatures, because as was mentioned before very, very few people really have an awareness of what this threat is. Thank you.

STOCKTON: Well, bless you for what you're doing, because so much authority and responsibility does lie at the state level. And a few weeks ago I went to the national meeting of the National Association of Regulated Utility Commissioners, NARUC. Again, bad on me, I had never heard of NARUC until I started digging deeper into this problem a year, two years ago and understood that regulatory authority for a lot of what we care about lies at the state level, not the federal level. And so if we're serious about cost recovery mechanisms that will actually work for industry to build resilience against these new threat vectors, including cyber, we need to talk to utility commissioners about the importance from a national security perspective from a DOD mission assurance perspective of building resilience, of the capital investment that is necessary to do so on the part of utilities. Someone has got to pay for that. I'm not paying for it. DOD doesn't have money appropriated to us for that purpose. There has to be a commercial basis for making this progress, and I'm all about working with the states and utility commissioners to ensure that they understand our perspective on why building resilience against these kinds of threat vectors is so essential.

CHUCK: And our last question.

GEORGE BAKER: George Baker, and I was in to brief your Energy Security Working Group recently. I'm really pleased to see that you are actually pursuing building peaking plants on military bases. I think that's a great initiative. I wondered if you are also looking off the base at prioritizing the generation plants, the commercial generation plans and maybe helping us make the case for protecting, we can't protect them all, but protect the ones that are particularly important for national defense.

STOCKTON: I think that's an important opportunity. I also think that as we look at mission assurance building redundancy into our supply in an intelligent and informed way and not having single points of failure for mission execution, really a systems engineering and a network theory understanding of what it takes for the Department of Defense to serve the people of this nation, that's a critical component. And as you've suggested, thinking from a big picture about which sources of energy are going to be most important to be able to access and how to be able to adapt, have some resilience in case some go down, to be able to execute from a load management perspective, exactly what we need to be able to do and know more, that's a path I'd really enjoy walking down with you and everybody in this room. Thanks.

CHUCK: Assistant Secretary of Defense, Paul Stockton. On behalf of our group I'd like to thank you for your personal investment of yourself and your office, bringing your attention and the nation's attention to this issue. Thank you very much.

Securing the Grid: Decentralization and Distributed Power
Richard Andres, National Defense University

Throughout the last decade, a number of long-duration and widespread power outages in the United States have served as visible reminders that the electrical infrastructure on which the country depends is both fragile and vulnerable. The 2003 grid failure, the result of a sagging power line brushing an untrimmed tree branch, left 55 million people in North America without electricity for a week and cost over $6 billion. More recently, the June 2012 Derecho knocked out power to four million people between Ohio and Washington, DC while Hurricane Sandy left nearly five million customers without electricity—many of them for weeks and some for even longer.

Although events on a large scale remain rare, grid instability and outages are common and carry an economic cost of $120 billion annually. More worryingly, while blackouts faced to-date almost exclusively stem from natural disasters—predictable events with predictable consequences—a far more serious set of low-probability, high-consequence threats confront the power grid network.

The first set of nation-level grid security threats are posed by cyber attacks. Seeking efficient ways of managing power, the electric industry has moved to internet-connected SCADA systems and infrastructure. Highly susceptible to attack, these systems are probed daily by a wide range of state and non-state actors. While these intrusions have not yet resulted in any catastrophic attacks, security experts largely agree that a number of hostile actors are capable of shutting down portions of the grid.

Second, the grid is vulnerable to a range of physical threats beyond natural disasters. This includes small explosives and high-powered handheld weapons. Similar to a coordinated cyber attack, a synchronized physical assault on multiple transformers across the country could result in long-term outages.

Lastly, an EMP event occurring from either a geomagnetic storm or high-altitude nuclear detonation represents the greatest threat to the U.S. power grid. Although these two occurrences are distinctly different, the risks they pose are largely the same. Geomagnetic storms, caused by solar flares emitted from the sun can induce extreme electrical currents in high-voltage transmission lines connecting bulk-power systems throughout the country. These voltages can overload and burn out transformers, permanently damaging infrastructure. An EMP originating from a nuclear detonation would likely have a far more intense effect and potential adversaries of the United States have publically described such attacks.

The U.S. power grid as currently conceived is the result of growth over 100 years and a centralized system relying on large power plants and an enormous distribution network of high-voltage transformers and transmission lines. This centralization and interconnectedness creates massive singular points of failure throughout the grid network—which can further lead to cascading outages—as the 2003 blackout demonstrated.

However, if the U.S. grid was modernized and decentralized—incorporating widespread integration of smaller power sources closer to end-users, including renewable energy options—the risk posed by the threats described above, in addition to natural disasters, would decline dramatically. Such a system was incorporated within Cuba during the past decade and led to increased grid resiliency and a decrease in the number of blackout days per year from nearly 200 to almost zero.

In 2003, Kodak's factory in New York escaped unscathed by having its own power units onsite. Ensuring that all critical institutions and infrastructures have a similar ability to generate power from sources available during crises is common sense. Furthermore, having local sources of power creates an "edge" capability that can be leveraged to restore the national grid in the event of collapse. Beyond this, such a system would go a long way to prevent grid failures completely as the production of power closer to end-users would reduce the need for high-voltage transformers and transmission lines.

Although obstacles exist to such an overhaul of the grid—including the government-granted monopoly enjoyed by utilities and the costs related to updating physical hardware—the system needs substantial reform as a matter of both economic and national security. The current centralized system is not only more inefficient and expensive, it is also significantly more vulnerable to a range of natural and man-made threats. A decentralized system will not make the grid invulnerable, however, it will go a long ways towards increasing the overall resiliency and reliability of the country's most critical infrastructure.

EMP Knots Untied: Some Common Misconceptions about Nuclear EMP
George H. Baker, Professor Emeritus, James Madison University

The author was originally approached to make a presentation to the 2012 DuPont Summit on the benefits of microgrids for mitigating the effects of electromagnetic pulse (EMP) and solar storms. On reflection, knowing the audience would be an exceptional mix of technical and policy leaders, he suggested that it might be more a propos to lay to rest misconceptions regarding the EMP phenomenon, its effects on systems, and the consequences of those effects. The author is grateful that this topic was accepted as part of the DuPont Summit agenda.

The author had no difficulty immediately listing a dozen misconceptions about EMP encountered during discussions with both technical and policy experts, in press reports, on preparedness websites, and even embedded in technical journals. Because many aspects of the EMP generation physics and its effects are obscure and non-intuitive, misconceptions are inevitable.

The wide-area, ubiquitous effects of EMP, and the numbers of systems potentially affected makes it convenient to adopt misconceptions that avoid the need for action. Denying the seriousness of the effect appears perfectly responsible to many stakeholder groups. On the other extreme, doomsday hyperbole is also present in some camps.

Misconceptions representing over- and under-emphasizing hyperbole have served to deter action in the past. Downplaying the threat places EMP preparedness on the back-burner compared to other effects. Exaggeration of the threat causes policymakers to dismiss arguments, ascribing them to the "chicken-little" syndrome.

Given the allotted 15-minute Summit presentation time, the author has limited the present discussion to his perceived highest priority misconceptions, or "EMP knots."

1. EMP will burn out every exposed electronic system.
2. EMP effects will be very limited and only result in "nuisance" effects in critical infrastructure systems.
3. Megaton class weapons are needed to cause any serious EMP effects—low yield, "entry-level" weapons do not engender serious EMP effects.
4. To protect our critical national infrastructure would cost a large fraction of the GNP.
5. Only late-time EMP (E3), not E1 will damage electric power grid transformers.
6. Long-haul fiber optic lines are invulnerable to EMP.
7. Ground burst EMP effects are limited to 2–5 km from a nuclear explosion where blast, thermal, and radiation effects dominate.

Misconception 1: EMP will burn out every exposed electronic system

Based on DoD and Congressional EMP Commission's EMP test databases we know that smaller self-contained systems that are not connected to long-lines tend not to be affected by EMP fields. Examples of such systems include vehicles, hand-held radios, and unconnected portable generators. If there is an effect on these systems, it is more often temporary upset rather than component burnout.

On the other hand, threat-level EMP testing also reveals that systems connected to long lines are highly vulnerable to component damage, necessitating repair or replacement. The strength of EMP fields is measured in volts per meter. Thus, to first order, the longer the line, the more EMP energy will be coupled into the system and the higher the probability of EMP damage. Because of their organic long lines, the electrical power grid network and long-haul landline communication systems are almost certain to experience component damage when exposed to EMP with cascading effects to most other (dependent) infrastructure systems.

Misconception 2: EMP effects will be very limited and cause only easily recoverable "nuisance"-type effects in critical infrastructure systems

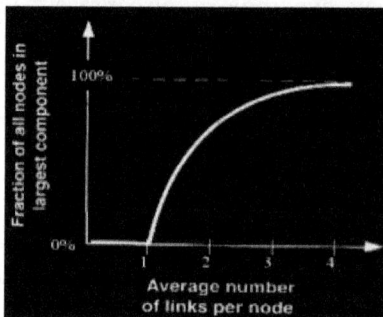

Although EMP does not affect every system, widespread failure of a limited numbers of systems will cause large-scale cascading failures of critical infrastructure systems and system networks, because the interdependencies of the non-failed subsystems with the electronic systems debilitated by EMP. Paul Erdos' "small world" network theory applies.[1] The graph on the left illustrates that the fraction of nodes in any network that are connected to single network node changes suddenly when the average number of links per node exceeds one. For example, a failed node, where the average links per node is 2, can affect ~ 50% of the remaining network nodes.

Also, for many systems, especially unmanned systems, upset is tantamount to permanent damage and may cause permanent damage due to control failures. Examples include:

- Lockup of long-haul communication repeaters
- Upset of remote pipeline pressure control SCADA systems
- Upset of generator controls in electric power plants
- Upset of machine process controllers in manuf. Plants

---

[1] Duncan Watts, *Six Degrees: The Science of the Connected Age* (2004).

Misconception 3: Megaton-class nuclear weapons are required to cause serious EMP effects. "Entry-level," kiloton-class weapons won't produce serious effects

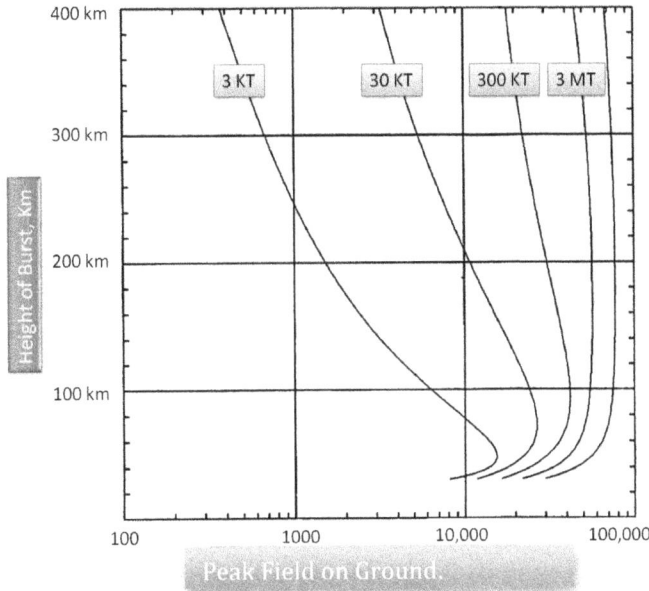

Due to a limiting atmospheric saturation effect in the EMP generation process, low yield weapons produce peak E1 fields of the same order of magnitude as large yield weapons if they are detonated at altitudes in the 50–80 km range. The advantage of high yield weapons is that their field on the ground is attenuated less significantly at larger heights of burst (that expose larger areas of the Earth's surface).

The first graph above illustrates that nominal weapons with yields ranging from 3 kT to 3 MT (a three orders of magnitude difference in the yield), exhibit a range of peak E1 fields on the ground of only a factor of ~3, viz. 15–50 kV/meter.

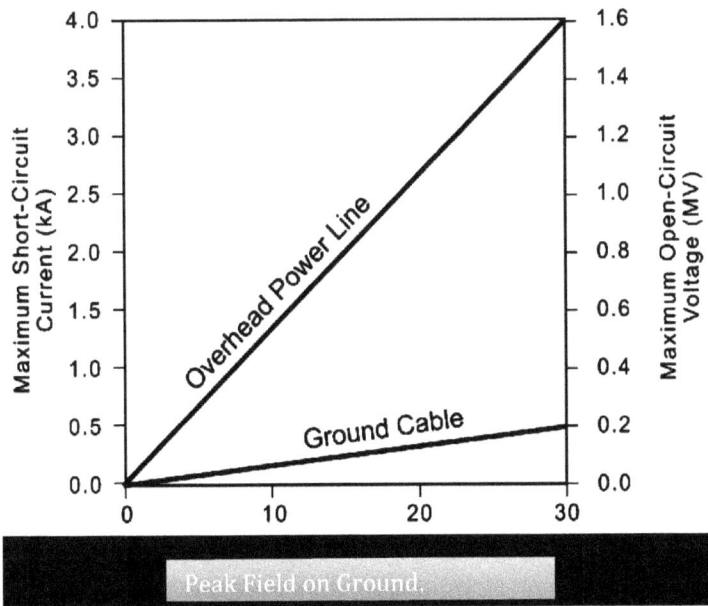

With respect to the late time (E3) EMP field, a 30 kT nuclear weapon above 100 km causes geomagnetic disturbances as large as solar superstorms, but over smaller regions.

The second graph on the left indicates that megavolt levels and kilovolt-level currents are induced in long overhead lines by E1 from kiloton-class weapons.

Misconception 4: to protect our critical national infrastructure would cost a large fraction of the U.S. gross national product

765 KV Generator Step-Up Transformer

Of the 14 critical infrastructure sectors, EMP risk is highest for electric power grid and telecommunications grid—attention to these infrastructures <u>alone</u> would bring major benefits to national resiliency. These infrastructures are the most vulnerable due to their organic long lines. They are also the most critical to the operation and recovery of the other critical infrastructure sectors. It is ironic that our most vulnerable infrastructures are also the most vulnerable to EMP.

If we have to pick one infrastructure to protect, the top choice would be the electric power grid. Grid operational behavior is binary—it fails fast and hard over large regions disabling most other critical infrastructures. The grid is the most essential infrastructure for sustaining population life-support services.

Some major grid components take months to replace—years if large numbers are damaged. The primary example is high-voltage transformers (an example unit is pictured in the figure immediately above) which are known to irreparably fail during major solar storms and are thus likely to fail during an EMP event. Protection of these large transformers will buy valuable time in restoring the grid and the life-support services it enables.

The unit cost for HV transformer protection is estimated to be $250,000. The total number of susceptible units range from 300 to 3000 (further assessment is required to establish an exact number). The requirement and cost for generator facility protection are still undetermined but are likely to be in the same ballpark as transformer protection costs. The need for SCADA system protection is moderated by the ready availability of replacement parts and the relative ease of repair. Doing the math, the protection costs for heavy-duty grid components are in the single digit billions of dollars—a small fraction of the value of losses should they fail. Amortized over 20 years, the protection costs amount to pennies per month for electricity consumers.

Misconception 5: Only late-time EMP (E3), not E1, will damage electric power grid transformers

Oak Ridge National Laboratories (ORNL) E1 tests of 7.2 kV distribution transformers produced permanent damage to transformer windings in seven of the 20 units tested. The failures were due to winding damage including turn-to-turn flashover and primary-to-secondary flashover. The results are summarized in the table below.[2]

As an important side-note, transformers with direct-mounted lightning surge arrestors were not damaged during the tests. Similar tests of HV transformers are needed.

XFMR	Shots (#@kV)	Peak Voltage (kV)	Time to Peak (ns)	Surge Arrestor Present?	Notes	Result
ZS1						Pulser Calibration
ZS2	1@400	264	618	No	(1)	T-T failure
ZS3	2@400	288	668	No	(2)	HV-LV failure
ZS4	2@400	280	600	No	(1)	L-L failure
ZS5	1@400	272	550	No	(2)	HV-LV failure
ZS6	2@400	290	643	No	(1)	No damage
ZV1	1@400	296	601	No	(1)	No damage
ZV2	1@400	304	592	No	(2)	HV-LV failure
ZV3	2@400	110	100	Yes	(3)	No damage
ZV4	2@500	110	100	Yes	(3)	No damage
ZV4	2@780	116	110	Yes	(3)	No damage
XV1	1@400	212	500	No	(2)	HV·LV failure
XV2	2@400	115	110	Yes	(3)	No damage
ZW1	2@400	292	552	No	(1)	No damage
ZW2	2@400	16	Oscillatory	No	(4)	No damage
ZW3	2@780	100	110	Yes	(3)	No damage
ZW4	2@1000	112	105	Yes	(3)	No damage
ZD1	2@400	120	550	No	(5)	No damage
ZD2	2@400	20	Oscillatory	No	(4)	No damage
ZE1	2@1000	95	100	Yes	(6)	No damage
ZE2	6@780	95	100	Yes	(6)	No damage

Notes:
(1) External flashover on HV bushing: T-T failure denotes turn-to-turn failure; L-L failure denotes line-to-line failure

---

[2] W. Radasky et al. *The Early-Time (E1) High-Altitude Electromagnetic Pulse (HEMP) and Its Impact on the U.S. Power Grid, Meta-R-320*, Oak Ridge National Laboratories.

(2) No external flashover; HV-LV failure denotes a high voltage winding flashover to the low-voltage winding

(3) Surge arrester operation and no external flashover

(4) Surge applied to the low-voltage bushings with no external flashover

(5) Surge applied common mode to both HV bushings with external flashover

(6) Surge applied common mode to both bushings and both arresters operated

Misconception 6: Optical fiber networks are not susceptible to EMP effects.

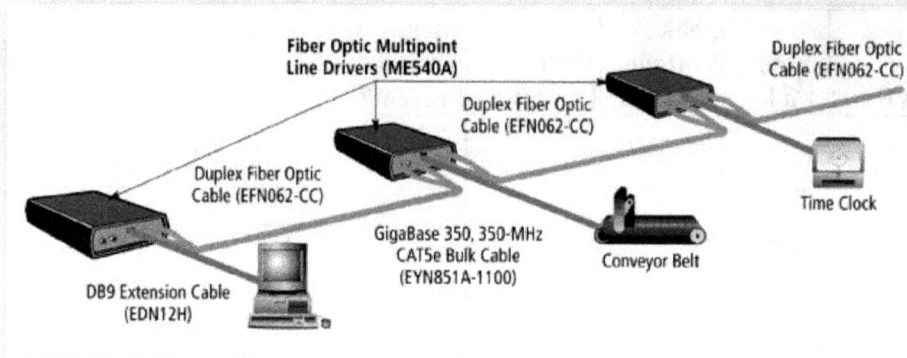

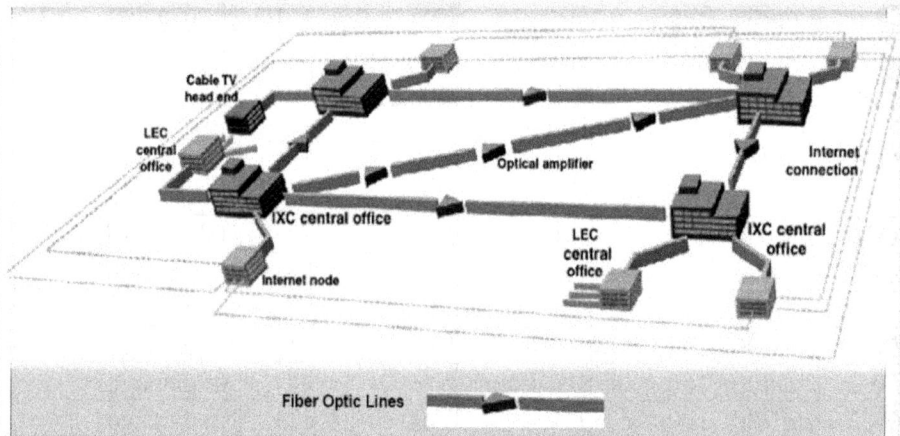

In general optical fiber networks are *less* susceptible than metallic line networks; however, fiber optic line driver and receiver boxes may fail in EMP/E1 environments. Long-haul telecom and internet optical fiber repeaters' power supplies are particularly vulnerable. Terrestrial fiber-optic cable repeater amplifier power is provided by the electric power grid and thus vulnerable to grid failure as well as direct EMP/E1 effects. Undersea cable repeater amplifiers are also vulnerable to EMP/E3 effects since, because of its low-frequency content, E3 penetrates to large ocean depths.

On the plus side, line drivers/receivers and repeaters are relatively easy to protect using shielding, aperture treatment, and power line filters and/or breakers.

Misconception 7: Ground burst EMP effects are limited to 2–5 kilometers from a nuclear explosion in the region where blast, thermal and radiation effects dominate. Thus, ground burst EMP is not a major threat

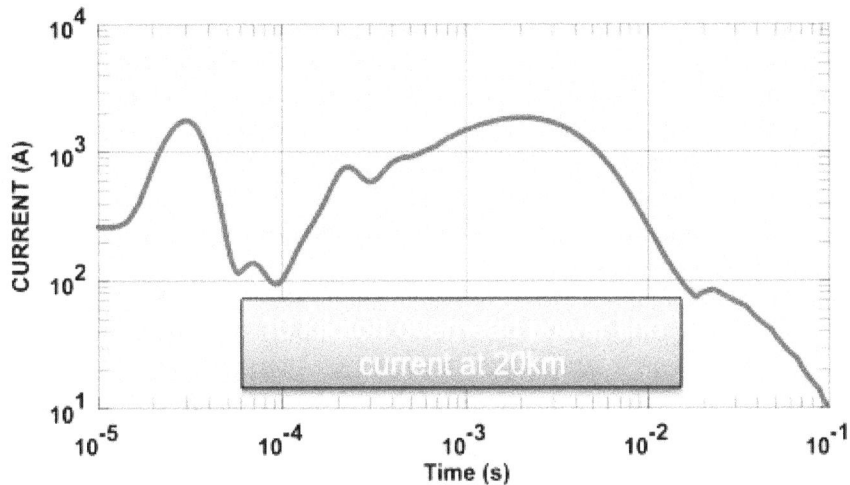

Ground bursts couple large currents to long lines running through the nuclear source region. These currents propagate to distances of tens of kilometers from the burst location. Destructive source region EMP (SREMP) effects on power and communications infrastructure extend significantly beyond the blast, thermal, and radiation effects ranges. As shown in the figure, a nominal 10 kT yield ground burst delivers a 2,000 amp pulse lasting for several milliseconds on overhead power line at 20 km. A 1 MT ground burst would deliver 150,000 amps at the same distance down the line. Long-line currents induced by a single burst can debilitate long-line communication and electric power networks over the area of a large city.

Conclusion

The seven EMP "knots" addressed here are common misconceptions and arguably the most important to "untie." There are others that should be addressed, but the present points, especially the first four, are crucial to dispel because they have deterred efforts to achieve national preparedness.

From a risk-based priority standpoint, the electric power grid is at the top of the list for EMP protection.[3] Hardening this infrastructure alone would have major benefits for national resiliency, i.e., the ability to sustain, reconstitute, and restart critical services. It is not just the survivability of our electric power infrastructure that is at stake; almost all of our critical infrastructure services will cease should the power grid fail.

A major impediment to action has been that government and industry are (understandably) swayed by the familiar, the convenient, and the bottom line. Like it or not, familiarity and profitability are the touchstones of acceptability—strategic

---

[3]G. H. Baker, . "Risk-Based Critical Infrastructure Priorities for EMP and Solar Storms," *Security Analysis and Risk Management Association Newsletter,*October 2011

advantage goes to the acceptable. Thus the tendency exists to downplay the likelihood of an EMP scenario and its associated consequences (Misconception 2).

By way of encouragement, we know how to protect systems against EMP. EMP engineering solutions have been implemented and standardized by DoD on a host of systems. In the case of the national power grid, the installation of blocking devices in the neutral-to-ground conductors of large electrical distribution transformers will significantly reduce the probability of damage from slow E3 component of EMP and geomagnetic disturbances (GMDs) caused by solar storms. Transformer protection against E1 overvoltages is achievable by installing common metal-oxide varistors (MOVs) on transformers from each phase to ground. Costs for protecting the power grid are a micro-fraction of the value of the systems and services and risk.

EMP protection methods for communication and control facilities have been developed and implemented by DoD since the 1960s and are well documented (ref. MIL-STD-188-125-1, MIL-STD-188-125-2, MIL-HDBK-423). Engineering approaches include use of shielded enclosures, provision of backup power, standard grounding techniques, installation of overvoltage protection devices and filters on penetrating conductors, and good cable management procedures.

Hopefully, this attempt to redress important and pervasive misconceptions concerning EMP will help to spur action on the challenging effects of EMP and public–private cooperation will begin and prevail in implementing low-risk EMP protection of our most critical infrastructure systems.

Cyber Security—Reducing Supply Chain Risk
Thomas R. Goldberg, Lineage Technologies, LLC

Demonstrable and validated (trusted) *state-of-the-art* commercial-off-the-shelf (COTS) computing and networking hardware can be manufactured and assembled in the United States from U.S. manufactured components and systems to meet Government (DODI 5200.44) and emerging critical infrastructure requirements. State-of-the-art components and systems, including servers, routers, switches, hubs, repeaters, and mass storage devices made exclusively from certified and traceable U.S. designed, manufactured, and assembled components, will significantly reduce the potential for the insertion of Backdoors, Trojan Horses, Trapdoors, and other elements of taint that permeate IT systems that threaten the United States today.

It is possible to significantly reduce the potential for deliberate insertion of exploitable vulnerabilities and malicious intent that can be used to sabotage operations and/or to steal vital information and data by shifting production to U.S.-owned and based operations, and by *redesigning, assembling, and delivering a state-of-the-art computing and networking equipment made exclusively of U.S. designed and manufactured components and systems.* All threats, including insider threats, cannot be fully eliminated, but by shifting manufacture and assembly, and authentication to the United States, the principal threat confronting the Country today can be significantly reduced.

The threat posed to the United States from tainted IT systems is so severe that entire weapons systems, banking and financial systems, and critical infrastructure sectors (natural gas, electricity, banking, etc.) have been compromised, suborned, and/or manipulated. The depth of the threat keeps increasing. Headlines detail attacks around the globe. On October 11, 2012, Secretary of Defense Leon Panetta again reported that the United States was facing the possibility of a "cyber Pearl Harbor" and was increasingly vulnerable to foreign hackers who could dismantle the Nation's power grid, transportation system, financial networks, and government.

Defining the Threat
In February 2012, Aviation Week and Space Technology[1] reported that in 2009 Chinese engineers were discovered to have participated in classified program/progress review meetings for the F-35, and had done so consistently for several years. In June 2012 NASDAQ's Chief Information Security Officer testified before the House Committee on Finance, and announced that "they are in our system, and we cannot get them out without direct support [of the U.S. Government]."[2] In the spring of 2012, the Administration announced that "nation-states" had manipulated systems controlling the transport of natural gas in the United States and Canada. This was confirmed and expanded upon by the Utilities Telecom Council October 2012. In early July 2012, Cyber Command told Senators that due to IT penetrations the direct cash outflows from U.S. banks in 2011 exceeded $338 billion, and that IP thefts for that period exceeded

---

[1] David Fulghum, Bill Sweetman, Amy Butler, "Budget Cyber Threat F-35 and Classified Programs Victimized by Network Intrusions," *Aviation.*

[2] Testimony of Mr. Mark Graff/CISO NASDAQ OMX before U.S. House of Representatives, Committee on Financial Services—"Cyber Threats to Capital Markets and Corporate Accounts", June 1, 2012.

$225 [3] billion.[4] When coupled with military/intelligence losses for 2010 and 2011 the estimated losses equaled $1 trillion in each of those years. In September 2012 Telvent, which sells supervisory control and data acquisition (SCADA) systems used to control critical functions at 60% of U.S. energy producers and transmission companies, announced that its systems were penetrated making them vulnerable to attack.[5] In October 2011, Verizon, U.S. Secret Service, and the Dutch High-Tech Crime Unit reported: "81% of all penetrations were installed, or injected at the point of manufacture"[6].

The principal attack vector involves the manufacture of computing and networking equipment that includes *"burned-in/on-silicon"* Backdoors, Trapdoors, and Trojan Horses that allow assailants unobstructed access to the operations and content of each device, enabling them to subvert or ignore firewalls, encryption, and most other protection systems.

Security metrology does not exist to completely verify the safety of IT systems manufactured abroad. This has been a normative fact since 1984 when first expressed by Ken Thompson in his Turing Award acceptance speech[7]. It was acknowledged again in House Select Committee on Intelligence in its October 8, 2012 Report:

"The task of finding and eliminating every significant vulnerability from a complex product is monumental. If we also consider flaws intentionally inserted by a determined and clever insider, the task becomes virtually impossible. While there is a large body of literature describing techniques for finding latent vulnerabilities in hardware and software systems, no such technique claims the ability to find all such vulnerabilities in a pre-existing system. Techniques do exist that can prove a system implementation matches a design which has been formally verified to be free of certain types of flaws. However, such formal techniques must be incorporated throughout the design and development process to be effective. They cannot currently be applied to a finished product of significant size or complexity. Even when embedded into a design and development process, formal techniques of this type do not yet scale to the size of complete commercial telecommunication systems... If significant flaws remain in widely fielded products and processes that are known to a potential adversary, it seems like the evaluation process has provided only marginal benefit."[8]

"...the United States' telecommunications sector increasingly relies on a global supply chain for

---

[3] "Energy Technology Provider Hacked by Chinese Group.", *Washington Post*, September 28, 2012, section A-17.
[4] Anthony Kimery, "Lieberman Calls for Cybersecurity Legislation in Response to Alarm by DOD's Top Cyber Chief", *Homeland Security Today*, July 11, 2012.
[5] Op. cit., *Washington Post*, September 28, 2012, section A-17.
[6] Wade Baker, Alexander Hutton, C. David Hylender, Joseph Pamula, Ph.D., Christopher Porter, Marc Spitler, et al., "2011 Data Breach Investigations Report," Report Prepared by Verizon/U.S. Secret Service/Dutch High Tech Crime Unit, November 2011, 27.

[7] Communication of the ACM, Vol. 27, No. 8, August 1984: 761-3.[8] Investigative Report on the U.S. National Security Issues Posed by Chinese Telecommunications Companies Huawei and ZTE, Report by Chairman Mike Rogers and Ranking Member C.A. Dutch Ruppersberger of the U.S. House of Representatives, Permanent Select Committee on Intelligence (October 8, 2012): 6.
[8] Investigative Report on the U.S. National Security Issues Posed by Chinese Telecommunications Companies Huawei and ZTE, Report by Chairman Mike Rogers and Ranking Member C.A. Dutch Ruppersberger of the U.S. House of Representatives, Permanent Select Committee on Intelligence (October 8, 2012): 6.

the production and delivery of equipment and services. That reliance presents significant risks that other individuals or entities—including those backed by foreign governments—can and will exploit and undermine the reliability of the networks. Better understanding the supply-chain risks we face is vital if we are to protect the security and functionality of our networks and if we are to guard against national security and economic threats to those networks."[9]

As the geometry of transistors on chips gets smaller (34, 22, 14 and 7nm) our ability to use test methods to determine security will diminish further. This was given currency earlier this year when Cambridge University researchers, Sergei Skorobogatov and Christopher Woods demonstrated that burned-in/on-silicon threats could not be found using currently deployed toolsets.

They employed a new tool to find keys on chips made for Actel/Microsemi by Taiwan Semiconductor Manufacturing Corporation (TSMC). The keys they found were on the ProASIC3 chips they examined:

"...would allow an attacker to disable all the security on the chip, reprogram any cryptography and access keys, modify low-level silicon features, access unencrypted configuration bit-stream, or permanently damage the device".[10]

They concluded the device was wide open to intellectual property theft, fraud, re-programming, etc. They also concluded that the penetration enabled attackers to reverse engineer the design of the chip thereby facilitating the design and introduction of newer versions of Backdoors or Trojan Horses. Lastly, they concluded that these penetrations could not be reversed or patched leaving every user vulnerable to unobstructed attack.[11]

Their discovery calls into question whether the use of functional verification, physical verification, and postmanufacture validation/debug can be employed for security purposes.

Solution
By relying upon U.S. designed and manufactured components and assembling IT systems in the United States, and by controlling access to those items during their manufacture, assembly, and transport, the principal threat vector affecting U.S. IT systems can be significantly reduced, if not eliminated. This outcome is being sought by new regulations emerging from DOD. They involve compliance with section 818 of the 2012 National Defense Authorization Act (NDAA) as set forth in DODI 5200.44, (11/5/12).

There has been great skepticism in the IT community that compliance with 5200.44 can be achieved, and that the systems of control (track/trace; tagging-tracking-locating, etc.) can be employed to build confidence. This skepticism is based on the belief that the current supply chain is too severely fragmented and opaque to be auditable. Attempts to measure the capability and capacity of the supply chain made this year by IHS/iSuppli and IPC yielded only limited

---

[9] Ibid., 7
[10] Sergei Skorobogatov, Christopher Woods, "Breakthrough Silicon Scanning Discovers Backdoor in Military Chip". http://www.cl.cam.ac.uk/~sps32/ches2012-backdoor.pdf: 1
[11] Ibid., 1.

data. These results reflect commercial databases that either track potential investment decisions, or parts by design and manufacturing company. These datasets lack the fidelity necessary to identify the site of production, or the capability and capacity to produce individual components. A recent classified Rand Corporation report concluded that the supply chain was not well enough organized to respond. Thus, what information there is on this matter is not encouraging.

That notwithstanding, confidence in this arena should be high due to two factors: (1) demand from IT integrators and principal sub-tier suppliers seeking all-American systems; and (2) the emergence of a quiet revolution wherein U.S. IT firms are returning production to the United States in order to protect their IP from expropriation and/or theft.

While Defense Advanced Research Projects Agency (DARPA) makes investments in new technologies aimed at re-engineering chip and software architectures as a principal means of overcoming present day threats, there is an immediate need for secure COTS products that are manufactured in the U.S. DARPA's undertaking will take several years to realize, so the interim solution of U.S. manufacture is required to close the door adversaries are using to fleece the United States and other developed nations.

Market Hurdles
The October 2012 House Permanent Select Committee on Intelligence Report provided an unambiguous voice to the U.S. Government's anxiety over foreign sourced computing and networking systems by pointing out that Huawei and ZTE products provided the Chinese military unfettered access to networks and systems wherever they are deployed. The Committee also hinted at the fact that U.S. branded computing and networking products incorporate components manufactured by these two companies and their suppliers. Thus, the Committee concluded that American computing and telecommunication systems are vulnerable to penetrations, IP theft, and subornment.

The tremendous savings derived from outsourcing over the course of the last two decades depended on de-leveraging manufacturing assets. This process created huge profits for U.S.-based firms. Unfortunately, today the result of those decisions has led to huge cyber losses. Those decisions now affect thousands of industries and enterprises whose IT infrastructure is wide open to exploitation.

Complicating matters further is the tremendous cost for re-establishing manufacturing in the United States, estimated to be in the hundreds of billions of dollars. These calculations are made more complex by the fears of the branded product producers that repatriation of manufacturing to the United States would imperil market access in China, the biggest market they currently have.

The least understood, yet possibly the greatest disincentive for repatriating manufacturing involves product liability risk resulting from embedded threats that have existed in IT products and systems for years. The capital and IP losses reported by Cyber Command to the Senate in July 2012, coupled with those of individual consumers, constitutes a potential liability exposure that could cost the computing and network equipment industry trillions of dollars. A decision to repatriate production to the United States to overcome embedded cyber threats could expose branded products to liabilities greater than those faced by U.S. tobacco companies during their fight over cancer causation.

The cost for redesigning/rebuilding the entire U.S. IT infrastructure will be enormous, costing trillions of dollars. This cost may be too great to be assumed in the near term and will significantly affect the bottom lines for all computing and network system providers.

Policy Hurdles

Among the most difficult policy issues to resolve in combating cyber attackers has been the need to secure the privacy of individual citizens. The threat to confidentiality, and in particular its aggregation in U.S. Government hands poses the biggest policy hurdle for lawmakers worried about Big Brother. Protecting the confidentiality of information in the United States while combating cyber attacks from abroad makes the task for U.S. cyber warriors almost impossible to implement. They can look at all IT traffic abroad, but dare not tread on traffic here without court-ordered wiretaps.

Our most cherished possession, our freedom, some argue, would not withstand the prying eyes of the U.S. Government into corporate/private databases. It will take time before these beliefs change, so quick action in the legislative arena is not likely to occur.

Yet despite these limitations, the United States may have little choice in deciding to aggressively defend the U.S. cyberspace. Each day gives new evidence to ever increasing losses and greater threats to our economy and our safety.

Administration Policy

Presidential Policy Directive (PPD) 20 was issued October 2012 and sets out the policy and procedures for interdicting cyber attacks against the United States. The policy allows attacks by DOD on systems linked even remotely to attackers against the UnitedStates. It also allows DOD to disable, damage, and destroy IT systems that are even remotely associated with attackers, including those of the target of an attack, if such an attack could have catastrophic outcomes.

This policy sets forth offensive and defensive doctrine for dealing with attacks on the U.S. mainland and U.S. bases abroad and reflects the determination of the Government to prevent loss of life.

A less well-developed policy is circulating in the form of a draft Executive Order entitled "Improving Critical Infrastructure Cybersecurity". This draft Directive seeks to fill the vacuum created by Congress' inability to pass new cyber security legislation. It seeks voluntary compliance with strategies developed and implemented by Federal Departments or Agencies under their existing statutory authorities. It seeks to provide the following framework:
- Information sharing regarding unclassified cyber threats to the homeland that identify specific targets.
- Establish mechanisms whereby the Federal Government can assist owners and operators of critical infrastructure to protect their systems from unauthorized access, exploitation, or harm.
- Protect privacy and civil liberties.
- Establish a baseline framework for reducing cyber risk to critical infrastructure

components through standards, methodologies, procedures, and processes that align policy, business. and technological approaches to cyber security.
- Modifying Federal procurement standards and guidelines to create preferences for vendors who meet cyber security standards.
- Use existing statutory authority to ensure the continued safety, environmental protection, and preservation of human health.

The draft EO directs Departments and Agencies to use their existing authorities to require computing and network system security in permits governing operations at any number of industries, including chemical, petroleum, power and related plants throughout the country.

Existing authority can be used by departments and agencies to open permits to require specific security standards for computing and network systems. Thus, by way of example, EPA regulations that already govern process and pollution control equipment can now require that SCADA and network equipment meet new and tougher security standards. FDA could require greater security in the network processes used to make, package, transport, and distribute proprietary and over-the-counter pharmaceuticals. DOT can require the same of pipeline operators, and hazardous materials transporters whose tagging/tracking/locating systems could be vulnerable to attack.

The examples of new regulatory provisions are endless. DOD has gone furthest with the issuance of DODI 5200.44 that regulates the supply chain for any electronic component and system sold to the Department of Defense. This Instruction, issued in November 2012, requires that each element of the Department modify its procurement practices to protect mission critical functions and to achieve trusted systems and networks. This effort aims to advance the state-of-the-art in assurance tools, techniques, and methods for creating and identifying software and hardware that is free from exploitable vulnerabilities and malicious intent. It is this policy that establishes the demand function for domestic design and manufacture of computing and network systems, and will likely lead to the repatriation of these functions for COTS products to the United States. It is this policy that will likely serve as a framework for implementation of the draft EO that is pending for all other Departments and Agencies. Because PPD-20 assumes that a vulnerability, however remote, threatens all assets in a network, we can safely assume that every effort will be made by the Administration to secure the network environment throughout the nation, giving further impetus to the conclusion that domestic manufacture of computing and network systems is an inevitability.

The Cyber Executive Order
Thomas Goldberg, Principal, Lineage Technologies

Mr. Thomas Goldberg, *Principal of Lineage Technologies, comments on the difficulty Congress has had in creating cyber-security legislation and the emerging presidential executive orders that are likely to be used as stop-gap measures. He discusses various cyber-attacks on infrastructure including on power generator assets that damage hardware and kill people. Goldberg's most significant points are that code etched into chips at the microscopic level allow those who produce the chips to sell access to the computers using them in such a way that passwords and network security protocols are completely ineffective at protecting against intrusion. The only choices presented were to live with little to no network security, disconnect any system containing IP from the Internet or bring back chip manufacturing back to the US.*

THOMAS GOLDBERG: Thank you very much. I'll address the… thanks. It was nice, also, to just briefly talk about the Executive Orders, the Presidential Policy Directives - I don't have the slide to that, but I'll speak to that at the end of the presentation. Quickly, I want to bring you to the threats as defined in the headlines. Mike Hayden in 2008, of course, said 15% of our assets are secure in Defense and Intelligence, leaving the vast majority exposed. Just before retiring, of course, General Jones told executives from the critical infrastructure utility community principally to hang up the phone, in other words, disconnect from the Internet and then Keith Alexander, who now runs Cyber Command, but also runs NSA, has said in a number of speeches, some of which we helped to write, that, you know, look, this is the largest wealth transfer in all of human history that's going on through cyber networks that are connected around the world allowing assailants to just take money out of banks and to take IP out of companies and then to plant themselves in utilities, in distribution networks, in pipelines, and things of that nature. As we go down this list, you can see that the Chinese, who we consider one of the more sophisticated attackers in the system, basically have fifth generation aircraft capability that they obtained from us. You my not have heard this, but this has been in publication in the full and open for quite sometime, that we discovered in 2009 that for quite a number of years, Chinese intelligence officers were participating with Lockheed and their subcontractors in all program and progress review meetings. So they were part and parcel of all of the dialogue going on resolving issues technically or announcing the progress along the line, and today, for all intents and purposes, your F-35 is owned by China and all of their new aircraft, the J20, J31, CH4, UAVs, are, if not mirror images, they're so close they might as well be and they cost a lot less. So, in any case, you go down the list of the threats, NASDAQ testifying before Congress at the end of May, first of June, depends on which record you go to, it's either the 31st of May, first of June, the CISO who just came over from Sandia National Labs, and CISO is Chief Information Security Officer, testified before the House Banking Committee, saying they are in our systems, we cannot get them out unless NSA comes to our aid, in essence, but what he also testified to is there is no NASDAQ trade that has integrity any longer, since we don't control ___**(04:27).** He says we, that NASDAQ, no longer has the ability to authenticate and validate and verify each and every trade because of the penetrations made there. FTC says that about 9 million people a year lose their identity through identity-theft over the Internet.

Probably more important for industry, and you'll see this at the very, very bottom, this is the very, very, very good work of the FBI, InfraGard in particular, reaching out to companies. Dupont in December of last year, because of the penetration at its company, took all of its intellectual property off of all of its IT systems. They went back to a very old process of using Dupont couriers; many of you will have remembered in the James Bond movies where people had little briefcases handcuffed to their wrist and they would travel around the world, that is the process that Dupont has returned to. Dow lost a billion dollars worth of research and the company you may not know much about, but is in Buffalo, New York, Lubrizol, was the beneficiary, if you consider this a benefit, of a hacking through the SCADA Systems, that's a supervisory control mechanism that control processes, pressure, temperature, residency in a process that is a real gem of knowledge on how to do something in a process environment. Well, the Chinese had managed to get all of that data, they enticed a disgruntled engineer from Lubrizol to Vegas, they presented all the data to the engineer to get a reaction. The reaction was shock and horror, how did you get all this? Once they knew that the data they had was valid, they built a plant in South Korea, and in February of last year they called up Lubrizol and basically asked for a billion dollars in blackmail money, or they would run Lubrizol out of business by making the thermoplastic polyurethane, which is an aerospace grade polyurethane, and do it at a fraction of the cost. Well, Lubrizol bought that plant for a billion dollars, took a charge against its earnings. It's only, at the time, a $2.3 billion company, so that's half of its revenue. In November of last year, after negotiating with the South Korean government to tear down the plant, Warren Buffett bought it at $0.04 on the dollar, which means that all of the assets owned by the shareholders was lost. And these are real events, so when we look at the EMP moment, what I want you to think about by the time I get to the end of this talk is, if we have to correct some of the activities in this area, we will be doing service in the area of concern to you and EMP.

Now, let's go to a kinetic effect. We've heard about Stuxnet, but Stuxnet is a bit ephemeral when you're looking at, if you will, centrifuges going at 1000 times their normal rate of rotation. Let's look at a real kinetic effect. This is actually an event where the Georgian government thanking the Russian government for having a short war that moved a pipeline out of Georgia and back into Russia, and you may remember this from about ten years ago, bought one million administrator addresses from the Chinese, because administrator addresses reside on chips. These are burned in on the chip and they identify each and every chip. And then they pinged around the world, because the Chinese didn't know where all the chips went, and they pinged around until they found the one that they wanted and it happened to be in a new computer in a processor on a server in this power plant in Siberia, and they identified that one of the nine hydro turbines was in lockdown/tagout moment for repairs. Lockout/tagout is a procedure where if you don't do the startup properly, step one isn't done properly, you never even get entitlement to go to step two. But because the chip had an administrator address on it that gave all privilege to the person owning that address, they dialed up stage one of that turbine, drove it out of six stories of solid concrete and steel, killing 75 people and destroying the entire turbine house of this hydro-electric plant. So this is your kinetic moment. In the little yellow boxes there, if you can see that, that's the size of the human beings examining the destruction after the event, so that gives you a scale in terms of magnitude.

When we look at economic effects, the first, very, very first news conference that Christine Lagarde gave after taking over the IMF, had in its very first sentence, "Shame on you China, for manipulating the votes and minutes of secret meetings of the IMF to serve your own geopolitical interests". What happened were a couple of members of a particular meeting that had voted in a particular way on a decision had discovered that the minutes, when published, were very much different than they recalled from the meeting. So they went back to the recorder notes, which are these semi-cardboard papers that the recorder, the court reporter type person, generates, read an entirely different outcome than that which was published, and the reason was, again, the Chinese using chips that had administrator addresses in them that they had the privilege to, were able to drill all the way through everything that was protection at IMF, and I'll stress this as we go along, and basically they owned that access. So if you can imagine, if you will, you're Maxwell Smart, for those you who remember Get Smart, and you're at the end of that tunnel at the end of the day and the doors are closed on you. Well, this administrator address privilege moment provides where all the doors open up instead; there is no encryption, there is no firewall, there are no air gapping mechanisms if you have access in even the most remotest way to these things, that will deny you privilege.

Leon Panetta, I've already told you about NASDAQ, but Secretary Panetta in a speech, again before financiers in the utility industry not more than a month and a half ago, just basically said, we have to look out for a cyber Pearl Harbor because, in essence, everything that makes up the electronic networks that we rely on, including all of the PDAs and things that we all have, have these frailties to them and that state actors, you know, we call state actors anybody that we don't like, basically that's another country, uses these addresses, just as they sold them to the Georgian government to give to organized crime around the world, and you have to wonder, why is it the Russians, if we write a million lines of security code, for one thousand line code can break in? Well, they're not breaking in because they're smarter software writers; they're writing to the address on the chip, and so they don't need a lot of code to be able to get around the mechanisms that we currently employ to protect ourselves. These are national security concerns. Probably the biggest one that is not really enunciated here, but should be, is that every time somebody penetrates your system, it isn't what they take that's most important; it's what they left behind, which is basically that latent kill switch or that ability to destroy the asset that controls something or will leave behind an instruction that if it's targeted will cause some catastrophe or failure in your system and, depending on how your system operates and what it's connected to, again, you've heard all day about cascading events. And the number of attacks is constantly increasing. We've given up, in many quarters, trying to really count it, but put it in a log basis that daily it's increasing by that, and the reason is that the number of administrator address privileges seems to be, if you will, percolating more.

This slide was done by Verizon, the US Secret Service, Dutch High Tech Crime Unit, they've been doing it on and off for a couple years, this is a euphemistic slide. It's deliberately trying to send a signal to the government that we need help, but if you look at the really, really big circle, it says that, essentially, that 81% of my vulnerabilities are installed or injected by some remote actor, meaning the manufacturer who sold you the computer, or the chip fab that actually fabbed the chip, or the packaging facility that packaged the chip. And then as you go down, you basically look across malware virus and all of the other vectors we hear a lot about, and in reality they're infinitesimally small by comparison. The supply chain; it is very complex. We built it, in a manner of speaking, to advance profits. We did this beginning in the 1980s with new theories and business schools, finance schools, if you will, management schools, that basically said, if you deleverage the manufacturing asset, your expense goes down, you own the customer, you own the profit and, guess what, your profit will explode. And so what we first did was we began to give to second parties in the middle of the country the ability to make our computers, then we shipped them to Mexico, then we shipped them to China. Well, what do the Chinese do? They have their own enlightened self interest. So they began engineering on top of our engineered product and we got back many, many things that we now no longer can control. But the convoluted nature of this is if you're in the IT community, and I am in that community, when you buy a part you never really know if you're dealing with a manufacturer. You may be dealing with a broker, who is, he or she, may also be dealing with a broker, so the depth and the breadth of this is so complex that what we have to do is literally collapse this back in on itself to get even a chance of getting hold of who's really in that loop. The state of the market today; everything is outsourced. Virtually everything is outsourced. The size and complexity of it, as I just said, is so enormous that when the Rand Corporation just did a classified report for the Secretary of Defense, their conclusion was that the industry supplying the assets to the Department had no idea how deep their supply chain was and how interconnected it was. So they gave up on that.

We have been blessed and cursed by performance enhancement. Each one of us today lives in the Buck Rogers world imagined in the 1930s. It's real. It's here. And yet it's cost-dominant, which means we get it at almost no cost. It's the way that John D. Rockefeller began his business making kerosene; he gave away the lamp as long as you bought the oil. We get an awfully inexpensive device. It is highly functionalized. It is a screaming demon; if you're like me, I used to race cars, we love to see that thing log on and boot up as fast as possible, we can watch movies on a train, whatever it may be. But security is an afterthought. Now when we, and I was one of a dozen or so folks asked by the Secretary to go around the country almost two and a half years ago now to talk to the industry and say, you're all Americans. Do your patriotic duty and bring some of this manufacturing home. I was given good coffee, occasionally got an Otis Spunkmeyer cookie, then I was shown the door, and the reason is that the US market for integrated circuits, if we look at the US government, component of that is 3% of what a chip manufacturer has as sales; not large enough to make a dent. And if you're looking at a systems integrator, and the one that has the largest number of sales, and that's Cisco Systems, it's only 11% of revenues. So there is no driver here to really bring it back.

When we look at how we're going to implement the changes that I'll briefly describe now, which is an instruction that was issued by the Undersecretary for Acquisition Technology and Logistics and DOD's Chief Information Officer, it is called 52044, it's the supply chain integrity requirement. Every division, every office that has a procurement function or a design function in DOD now has to issue regulations that essentially require that if a vendor sells a product to the DOD, the entirety of that supply chain is understood and that security trumps cost and other concerns. And the reason we need to do that is we have to get to the designer, and not all designers reside in the United States; in fact, a lot of the code that is used to design a chip was invented in Germany. A lot of the coders are Russians, are Belarus Russians, or folks that are in the Philippines, and if you look at databases that are generated as information databases, if you look at the workforce, it's not in the United States entirely, more often not, than here. When you look at your operating systems, your packagers, your appliance assemblers, you're going to have to go back to some level of control that we have had in years past, we call it the Trusted Systems Program, TAPO if you're at NSA, or trusted through DMEA, which is the counterpart to TAPO that is within the Office of the Secretary of Defense. What are our control capabilities? I've listed them down here in the blue, is what we can actually do in the United States, because of the supply chain in the red, by virtue of sovereign rights for other nations, we will not be able to achieve. Onsite inspection will not be allowed and the most recently retired president of China so stated over 18 months ago. Lifecycle controls, because you can't imbed at those locations, means you can't tag it, track it or locate it, and vetting of all the supplies if they're in foreign nations is an impossibility because of sovereign issues. But that's where we have to go. So achieving a secure supply chain really has taken life, as of the fifth of November of this year, in 52044, and we have to have in that visibility, cradle-to-grave we have to do validation and certification, and that involves people, company, facility systems, processes and parts, and here's the kicker; if you sell an electronic component, a widget, a device or a system to DOD and it contains taint, taint is the deliberate manipulation on the chip or its counterfeit, which means you bought something that has a label on it to do something when it was really and older generation or not designed for that purpose, you're fully liable. So if the aircraft falls out of the sky, the missile fails to hit its target, you're liable for the whole systems cost, and that's the driver here. Now, none of this is feasible in the current global cost footprint, so we either push oversight overseas or we're going to have to return manufacture to the United States.

So that gets me to Presidential Policy Directive 20. It's a classified document, but I'm only going to tell you what the Washington Post published about it, so we'll limit authenticity and validation verification to what the Post believes it says. But let's assume that there's some credibility here, and that is that the President may interdict any attacker anywhere in the world through cyberspace at their point of origin for even remotely, if that can be targeted, which means that anybody's computer anywhere that might be the transmission component in the Internet, can be destroyed or disabled, and that includes in the United States, if you're the target, you're system, to prevent a catastrophic event from occurring. There's a lot of credibility to it. The second one is an executive order that's been meandering around, hasn't quite issued; I was hoping it was issue by the time I got here. I keep praying for it because I'd want to be able to talk to the specifics, so I'm going to talk a little more ephemerally to what I expect will be in this Executive Order coming out next week, I pray it comes out, and that is because we have legal authorities that are very limited in interdicting the private sector owned Internet, which is about 90% of everything that we use, we are going to see authorities emphasized, not enhanced. That is to say that if EPA has an authority to regulate emissions of air coming off of a plant and it has SCADA systems that are vulnerable that could be compromised to cause emissions from occurring, they're going to have to change the SCADA systems under the air permit, under their NPDS water permit, under their RCRA hazardous waste permits. The same goes with FDA in terms of the controls that exist on the manufacture of pharmaceuticals, at USDA with regard to meat packing, and down the line. So any Federal agency will be told to use their existing authority in the absence of new statutes, which I will explain in a minute are not likely to occur, to essentially require the repair, the replacement of vulnerable systems. Now, at the moment that that goes forward, the EMP community ought to basically say, you know, by the way, why don't you harden it against electromagnetic pulse, because we're going to tear out those systems. Now, the case in point, and I alluded to it very briefly, you may have seen it in the pipeline moment at the beginning, Telvent, a Spanish company that had installed SCADA systems throughout the

United States, Canada, Central and South America, about 60% of that infrastructure is all common because the industry got together and said, hey, we're gonna have to come to each others' rescue, we ought to know a little bit about each others' tools, right? So there was good intent for putting that in. They reported to all of their customers that all of their SCADA systems are penetrated. So they have to be torn out, they have to be replaced, they have to be hardened for cyber purposes, there's no reason why not, in that process, they shouldn't also get an EMP incentive, and that really requires for a community like this to go out and say it doesn't cost much to shield. I mean, you've got companies like ARC Technologies up in Amesbury, Mass, that does this for lots and lots of people, why the hell not give them a little more business to put their foo-foo dust in these components and it will be done. So, I'm going to just close with this and then take a few questions. This is the consequence if we fail to act. Thank you to the Toronto Post. I saw this in the drawing board section of the Post and I just had to get a copy of this, so this is one of my mantras now, so maybe one day I'll have a chest of medals showing that I fought this battle through. I'll be happy to take any questions.

MALE: The first question, from Doctor Clay Wilson, who has been a scholar researching this area for some number of years and does cyber security at UMUC.

CLAY WILSON: I want to ask you, you're getting into preemptive cyber attack, which seems to be an emerging policy for cyber security since we're starting to say, well, we pretty much have to give up on the defense part. I've been reading lately where researchers and hackers are being paid money because of a new global market for sales of zero-day exploits. The purchasers of these zero-day exploits are nation states, maybe some businesses and potentially non-government organizations, extremists who may be using proxies, so do you have an opinion about how it's possible to control the sale of zero-day exploits so that they don't just create a cyber arms race, or are we beyond that point now?

GOLDBERG: My opinion is we're well beyond that. Those of us who have been working in this area behind the scenes in various offices have seen this going back to the year 2000, Y2K. I was sitting in a skiff with a number of members of the administration and Bob Rubin was asked, so how much is it going to cost us in lost revenue because of the backdoors, Trojans and so forth that will be burned in because we have outsourced the ability to correct that two digit to a four digit moment, because we didn't have people living in the United States that could write C, COBAL, Fortran, and fix the problem. The answer was $4 trillion dollars. The first time I ever heard the number trillion and my jaw dropped. So we've been dealing with these kinds of issues, not specific to the zero-day moment, but these issues for a very long time. We just haven't paid much attention to enhancing our defensive systems and we really do, if you're going to take CrowdStrike, made up of a lot of former FBI agents who recently left this year to start that company, their advice to clients, the first order is tear out your hardware, tear out your software, start over. So that tends to be where it is, and in answer specific to your question, what will the Administration do, I think the Administration has chosen to pound its chest and basically say, offense is our best defense right now.

DR. CHRIS BECK: I noticed on your last slide, not the cartoon but the one before that, the last bullet either push complete oversight overseas or return manufacturing to the US; probably the former is not likely due to sovereignty issues, so you told an anecdote about going around to companies and asking them about bringing their manufacturing back, and you were shown the door. I'm just wondering if there were any analyses of costs differential that we would see if chip manufacturing was done domestically versus overseas and what the difference in the chip itself would cost and what the difference to an overall system would cost.

GOLDBERG: Very good question. It's very, very low, and we know this because the IP losses of the companies that make chip; if you look at, and this is a quiet revolution, so there's hope on the horizon, I really mean there's hope on the horizon. Apple just announced today they're bringing in assembly function back, the rationale given, all of our IP is being stolen, we need to come somewhere where we can protect our IP, so if Apple's going to move it back, you can imagine Intel's moving some of it back. We do know quite a number of people whose IP was stolen. I want to give you one case study, very briefly; the largest selling encryption chip, the Proasic3 made by Actel Microsemi, was fully penetrated at Taiwan Semiconductor Manufacturing Corporation and had been done so for years and they are our ally; 1 million plus backdoor keys found in there by Cambridge University in a paper that was published over the summer but was first released in April of this year. So when you think about the ubiquitous nature of the threat, it isn't just coming from people you might consider your adversaries; it might, in fact, be your friend.

MALE: Any analysis done, kind of the Dupont or Dow model of pulling stuff off the network, but from a SCADA perspective, instead of ripping everything out and redoing it, the reason that attackers can get in, it's on the network. Any analysis been performed of just pulling the systems off the network and just keeping them almost as an intranet within the corporation of the utility company itself?

GOLDBERG: There has, and it depends on the company. They're doing a number of different things, but the principal way of doing this is since the chips in the pumps, valves, and controllers for temperature, resonance, stir, whatever, have themselves been penetrated, they're being sanitized by being replaced. Now, the networks themselves are being revisited. The use of Bluetooth or RF connectivity for enhancement is being changed to very much hardline in a closed loop fashion in some cases. I don't think anyone's going to go back to the engineer or the process mechanic, and by the way, I grew up in the chemical industry when I was very young, and we actually did have people who turned valves on schedule and did read meters to determine if things were running properly, but I don't think we're going back to that, but you have to sanitize everything in the system. You can't just isolate the system, and we know that from this fact; how did we get Stuxnet into the centrifuges? Well, it was an air-gapped system, meaning that it didn't connect to the Internet at all and it was in a closed building and it was in a place where people had to have authority to go into. Well, we flooded the market with thumb-drives that basically had Stuxnet on it and eventually one got there. Why? Because an engineer, and this happens, this is the human frailty moment and we're all human, actually brought something into the system they shouldn't have. And that's the human factors component of this, though I won't go any further on that.

LISA WRIGHT: Hi, I'm Lisa Wright. You indicated that you'd get to the question of legislation compared to an Executive Order and I invite you to do that now, and I'll just add if we are in a world where Congress cannot legislate by statute to provide the stability of statutes, and we are in an Executive Order world where the authority lasts only as long as the current president, that opens another whole set of questions.

GOLDBERG: Thank you for that question. Here's the frailty in the system, and it is not Republican, it isn't Democrat, it's Civil Libertarian versus collaboration between governments and industry. The Administration, both the Bush Administration and the Obama Administration, have attempted to work with Congress to get legislation to enable us to share data across government and industry, and Civil Libertarians won't have it. The fear of Big Brother is so intense that in the Senate, in particular, you're going to get an endless number of filibusters. It simply is not a bill, in my estimation, that you can expect to come off the floor of the Senate for that reason. And those are very, very sincere people who believe in freedom and in all of those aspects of freedom that are attendant to not having Big Brother in your bedroom. The difficulty is, and this is the odd nature or, if you will, the inevitable part is that Colonel Jang, and that's a made up name, he is in your bedroom, but the Federal government of the United States will not be, and that is because of the connectivity of the world and the access that people have and the like. There is another provision here that is very contentious, which might be resolved, and certainly has been called for to be resolved by Secretary Panetta, and that is that if you look at the systems that you own today, the providers of those systems are liable to you for the losses you have sustained. If you're Citibank or Goldman Sachs or a utility or you're Lubrizol and all of your information was stolen, is there somebody at fault? So, there is an absolute drive by the industry made up of the hardware and software providers and those that employ them, the Internet service provider community and what have you, to be exonerated from those liabilities and until such time they, too, are opposing the enactment of legislation that would enable it. So, with regard to executive orders, executive orders, they can survive if the new administration going forward chooses to continue to enforce them, but that's the only mechanism legal remaining that is a Federal authority in this particular space. You should expect, beginning in the next year, maybe the year following, a lot of litigation that might force the courts to get involved. I don't know really where that would lead, but those are the mechanisms that are available there being employed. I think they're doing the best they can with what they have.

Energy Security
Honorable James R. Woolsey

The Honorable James R. Woolsey, *former Director of the CIA, provided extremely candid comments about the severity of the vulnerability of U.S. critical infrastructure, and the inability of most political and bureaucratic leaders to imagine how adversaries can relatively easily cause a nationwide long-term power outage. His presentation's title, "Energy and National Security: Protecting Our Ability to Use Electricity," bridged both the need to protect large critical infrastructure systems and the need to make and store electricity locally through the use of micro-grids. He provided a further foundation for inventors of local power generation technology and their security.*

CHUCK: So we've been hearing a back and forth over the course of the day, some very general problems and policy considerations and some very detailed technical presentations, and we're just doing a little bit of each so that when you think through the policy issues you'll have some basis from which to understand the scope of the problem and the opportunity to solve them. I'm going to ask Jeff Weiss to come back up again.

What's interesting as he introduces the honorable James Woolsey to us Jeff has had a background which many of you can appreciate, because if you're in information technology you're called an IT person, if you're in information and communications technology you're an IC person, but what's best is if you could be an information, communications, and energy technology person; then you're an ICE-T person and it's a lot more refreshing, and it gives you the big picture of what goes on. Well, what's interesting about Ambassador Woolsey too is his broad background, because when you've been impacted by many of these things you get the big picture which very often moves you to action more than only having the small picture, because you always assume, "Well somebody else has this covered."

But right now I'd like to bring Jeff up who will give us a moment to introduce our next speaker. Jeff Weiss who is a micro grid entrepreneur and very active in the BENS studies on micro grids and the military bases. Jeff.

JEFF WEISS: Thank you, Chuck. So I'm here to introduce the honorable R. James Woolsey who is he who needs no introduction. Jim asked me if I knew what he was doing since I was asked to introduce him, and for those of you who don't know this already Jim is no longer Director of Central Intelligence, so if you're here for that you're in the wrong room. He is no longer Ambassador to the negotiation on conventional armed forces in Europe. He is no longer Under Secretary of Defense. He is no longer General Counsel to the U.S. Senate Armed Services Committee. He is Chairman of Woolsey Partners and he is venture partner with Lux Capital. So if you're interested in energy and need funds broadly from a technology and deployment point of view Jim is your man; talk to him.

I'd like a show of hands. Who in the room loves oil? Who in the room loves those who we buy oil from? So, depending on your answers to those questions you have something to talk to Jim about. Jim has taught us all about our dependence on oil. There is a television program called *CSI*. Some people think it has another moniker. From some points of view it's named for Jim. Jim is connected strategic and invested, so with no further ado please meet Jim Woolsey.

R. JAMES WOOLSEY: Thanks, Jeff. Is this microphone on? It sounds like it may be. I'm honored to be asked to be with you today. My background is not technical. I had six years of Latin years ago when I should have been taking chemistry and physics, but so be it. I want to put one basic distinction in front of you in hopes that people will always understand that when they're dealing with the country's security and major systems they cannot look just at naturally occurring or naturally caused events. Terrorists are smarter than tree branches.

We had tree branches knock down two transformers during the storms a few months ago at our farm near Annapolis. Had that been planned it could well have taken out the electricity for a huge area of Eastern Maryland. As it was it was 25 homes on a country road, transformers were relatively small, they were available. It took a few days, 25 cherry pickers. 20, 25 utility engineers, work people knew the system from Arkansas actually. And we put it back together again.

It can have a bigger reason, the electric grid that is, our electricity system can have a bigger reason to go down. Some time back I believe in '03 a tree branch—tree branches are quite sneaky things—touched a power line in Cleveland, and after several hours of hacking around in confusion some 80 gigawatts went offline in 40 seconds, taking out electricity for some 50 million people in the United States and Canada, many of those parts of the United States and Canada out of electricity for well over a week.

We tried to adopt a rather familiar tactic at least to those of you who have children or grandchildren who watch *South Park*, namely to blame Canada. But in their polite way our Canadian friends pointed out to us that actually Cleveland is South of Lake Eerie, not North, and it was our power line and our transformer that went down and our tree branch.

Tree branches touching transformers is what I call a malignant problem. Malignant problems can sometimes be relatively small, although naturally occurring, not only malignancy in the sense of cancer but let's say ease of taking a system down because of its complexity, butterfly effect. A butterfly flutters his wings on one side of the globe, cascading change as a result of the complicated interaction of systems, tornado on the other side of the world. But malignant problems however difficult they may be, and our electricity grid is highly, highly susceptible to them, however difficult, however problematical they may be they are generally speaking, not always but generally speaking far, far easier to deal with than what I call malevolent problems, problems in which there is an intelligence on the other side who is trying to decide, figure out how to destroy your systems or worst of all how to kill as many of you as possible.

We have an electric grid—I feel like Henny Youngman, the comedian. Take the electricity grid, please. We have an electricity grid whose construction was begun in the late nineteenth century and in its own way it's a marvel, possibly the biggest and most complex machine in the world, and it generally works. But we have only had as we built our American infrastructure for electricity, we have only had over the century and a quarter about that we have had something like a grid, we have only had a very brief period of time just before and just after actually Pearl Harbor when we were worried, because of Japanese and German submarines off the East and West coast, we were worried about some type of intentional disruption or attack on American infrastructure.

Our whole infrastructure was really put together, because of the two oceans that surround us and shield us from many of the world's problems, our infrastructure was put together without on the whole a single thought being given to intentionally caused harm. Our infrastructure was put together for ease of access, for convenience, for engineering efficiency, but think about how to design it so that a terrorist can't take down half a dozen of just the right transformers which he knows about from the web and thereby take out a substantial share of the country's operations—nobody was thinking about that when they put the grid together and nobody generally has thought about those kinds of problems at all. They seem too crazy. They seem too paranoid or something.

But electricity has lots of potential enemies. The grid has potential enemies in a sense to its smooth operation. I want to touch on just three of them here. There are many experts in this audience who know much more about the engineering issues and the technical issues of this than I do. But first of all we have physical destruction. If a tree branch falls on a transformer as it did at the edge of our farm it may break up the transformer, it may need to be replaced, but it doesn't, the tree branch doesn't go after the transformer right at the edge of the National Security Agency or the Pentagon or others that may be so large they have no spares or if they do have spares the utility, figuring it didn't want to cause the terrorists too much trouble, would put the spare transformer right beside the operating transformer. And just to make it clear what you ought to shoot at with your armor piercing round they would put "Danger. Do not touch" every place there is some place you should shoot if you're a terrorist.

The grid is not designed for security against malevolent interference. It is as bad as systems get in that regard. One of the reasons it stays bad and vulnerable is because partially the utility financing system which puts a great penalty on any added cost of electricity to consumers. It's partly because of our liability system in which if you raise these issues with utility executives and they are ones that they have spent some time actually thinking about these issues at all, you will find that there is no real sense of wanting to even talk about these issues, much less write anything about them, because the general counsel has told you that if something bad happens and there are any emails around or any memoranda record of discussion it could mean big liabilities for the company, so don't talk about it. Don't even think about it.

There is a woeful blindness about the interest in potentially hostile countries in physical attacks on our grid. One of the big interconnect organizations—I won't name it—a few years ago had a very smart electrical engineer from a large country with a sophisticated military, let's say that, come to him and say, "I'm doing research on electric grids. I wonder if you could let me have the maps of your grid so I can see where all the transformers are and so forth." And the government relations person of the interconnect company picked up the phone and called the State Department, and the State Department said, "Sure. He's from this country that we want to get along with so let's be nice to him."

So he spent months verging toward a year going through all of the control systems very thoroughly, taking notes and so forth. And then finally this issue, somebody worried about it a little bit and it bobbed up to the top of one of our national regulatory commissions. And there was a call to this gentleman from the foreign country, and the regulatory commission individual said, "Well, we understand you've been working on this for months and months, and we're interested. We would like to compare notes. We're interested to see what you've learned and what your notes say." And the gentleman from the foreign country said, "Sure, absolutely. I'll come to Washington the day after tomorrow." They said, "Fine." And then the next day he flew back home.

The degree of naivety and lack of willingness to put yourselves in the shoes of those who would do malevolent harm—it's the only way to do it—figure out how you could create the biggest problem, then figure out how to defend against it, that whole sequence of thinking that is second nature to people in many parts of our military and the like as far as I've been able to tell is almost completely absent in the electric business, almost totally and completely absent. So, in addition to the malignant problem, the tree branch touching a power line, got to be thinking about the malevolent problem of terrorists or others with say physical attacks on the electric grid. And if you don't you come up with major national debates on things like whether or not to put transmission lines underground.

And the only subject being debated is expanse and so forth. If you say, "You know, if you go to the expense of putting them underground and you're doing something thereby about tree branches, but you're not doing a damn thing, spending a nickel on figuring out how to keep terrorists from taking out transformers or otherwise physically attacking the grid, is this really— Shouldn't you have a little broader debate?" And you get back basically deer in the headlights look.

The second way in which we could suffer very serious harm from intentional malevolent interference with the electric grid is of course hacking, cyber vulnerabilities. Back in the late '90s as the web was moving into greater and greater use and people were intrigued by the low cost of passing information across it, and we were at the same time trying to deal with Y2K, so there was a lot of openness and willingness to think about changing the control systems of things like the utility grid, people had a great idea. "Hey, let's save money by putting the SCADA systems, the control systems for the grid over the web, and then the communication will be cheap and it will be easy. And what could go wrong?"

Well, there is one example of the kind of thing that can go wrong with Internet, bless it, which was put together by some individuals who in equal measure were geniuses and are, the ones who are still here are geniuses, but who also lack a certain, several of them, a certain sense of suspicion or concern or security, thinking, "What could go wrong?" That chain of thinking led to some rather substantial difficulties.

Take the SIPRNet which is the secret-level Internet that has hundreds of thousands of people on it with secret level and below security clearances, but not the very highest security clearances, but several hundred thousand people. That system was penetrated by one teenage Private First Class who put a thumb drive in, downloaded hundreds of thousands of classified State

Department cables, gave them to Wikileaks, and as they were published all over the world the people died, people who had been candid with American Ambassadors, with senior officials in our government, and now showed up on Wikileaks for having said X to the American Ambassador. Some of them were doing what they thought was right for their country, but they were being more candid with the Americans than their country wanted, and in some of those countries they solved that problem by killing these people.

So, if one teenage Private First Class can get into one of our country's most secure networks modeled after the Internet, now tell me why anybody thinks that you can't get past the firewalls and the SCADA systems into the heart of our electric grid. My youngest son is a pretty good hacker. I'm not. But I've got to say I have in the several years I've been working on these issues never run into anyone who thinks that the SCADA systems as they are now, unprotected, operating over the web, are anything but a potential disaster as far as malevolent interference is concerned. And burying the transmission lines or other types of fixes that focus on essentially malignant problems will not deal with malevolent problems.

One can of course also and we should worry very much about electromagnetic pulse of several kinds. Set aside the solar events for a minute. Those are things that we should deal with, we should have the right types of transmission systems and the rest so that we don't end up with the same kind of problem from the sun that we could have from malevolent interference, but the sun at least is not controlled by say China or Russia or Hezbollah or whomever. The sun will do what it does when it does it. In the meantime we need to hedge of course against some types of malignant problems such as that, as well as malevolent ones.

But the malevolent ones with respect to electromagnetic pulse are likely at least in my mind to be an extraordinarily severe problem. Why do I say this? Are people going to be launching nuclear weapons everywhere causally in order to just cause trouble? For our electric grid, no, but the difficulty that is presented is that particularly as the number of nuclear powers begins to increase the circumstances in which something could happen accidentally by way of a launch increase probably at some exponential level.

PBS has a one-hour presentation that just was on television a few weeks ago called *The Man Who Saved The World* about in 1962 the Soviet commander of the four submarines off Cuba that had nuclear torpedoes, how he decided and enforced his decision against the two other people who had keys to launch, having decided to launch he held out and kept the launch from occurring. I guarantee watching that show is a bit bracing from the point of view of anyone who might be inclined to say, "Well, nothing like that could ever happen."

We also know from releases in the aftermath of the collapse of the Soviet Union back 20 years ago, we know that Castro was doing everything possible to urge the use by the Soviets of a nuclear weapon, and he did not care that Cuba would be destroyed if he thought the United States would be destroyed. This is all very well chronicled in the Soviet materials. And Castro is not even a religious fanatic; he's just a Communist fanatic.

So, we also have to look at circumstances which could arise because countries such as Venezuela or countries that could have a fishing boat that could contain a scud, some 40 countries have scuds, all you really need to do is get close enough to the United States to launch up several hundred kilometers. And you might think, "Well, surely we're able to shoot something like that down." No, we have no deployed systems that are designed to deal with anything in the ascent phase. They're designed to deal at midcourse and in terminal deal with a system that is shooting at a target on the land, but if your target is several hundred clicks up, because all you want to do is detonate in space in order to cause an EMP effect, you don't need to have accurate systems; you just need to be able to launch something that has the right nuclear characteristics, not absolutely simple but not rocket science in a sense.

And this whole vulnerability that can be caused and could manifest itself in launches from someplace like Venezuela from fishing boats with a scud in it the way the Iranians have practiced launching, all of these possibilities exist. It is not likely in some sense that there is a way you can put a number on it and we're sure the number is X rather than Y.

It's just an understanding that in the Mideast after Iran gets a nuclear weapon where that means probably Saudi Arabia, Egypt, Turkey, and maybe some other Middle Eastern nations will have nuclear weapons. The next crisis in the Middle East could well produce some use of nuclear weapons. If any of them happen to be in the Western Hemisphere fire it up several hundred kilometers, while still ascending being detonated, we could, I don't have to tell this group, lose a substantial share of our electric grid and of course since we have in the electric grid only one of the 18 critical infrastructures in the country, water, sewage, etc. but it is still the case that all of the others depend on electricity. So, as millions of people found out in the various storms and small hurricanes and so forth this past spring and summer and fall, one can end up losing electricity and then one finds that nothing else works.

How can we deal with some of these? There are lots of ways. You're going to be more involved in it than I. I tend to be very favorably disposed toward distributed generation, toward improved storage so that one can put generation capacity where the load is where possible, systems that can island easily, systems that can make it possible for electricity to be used in such a way that even if you can't have all of your electricity, even if the military base has its soldiers and airmen pretending they're in England and drinking warm beer and taking cold showers, nonetheless if the runway lights operate the task critical assets can function and the base can still conduct its mission. Same with hospitals, police stations, and so forth. There are ways to work on the possibility of keeping a bare bones at least of electricity generation, even if the major systems, transformers, transmission lines, etc. have gone down.

I don't offer any very specific solution here, but it is an issue and an area that in the investment community people are starting to look at very hard, how to mate up affordable and safe storage of energy, particularly electricity but not exclusively, but mainly let's say electricity in such a way that we can have at least 20 or 30 or 40% of our electricity if the grid goes down.

I think those three ought to be enough for a reasonable assignment, even to as knowledgeable and able a group as this. The key thing, however, is as far as I'm concerned not to let anyone get away with the proposition, "What could go wrong? The only thing that could go wrong is something that is natural. The sun will do what the sun does. Tree branches will do what tree branches do and they fall. And if we deal with that set of problems, for the tree branches we bury the transmission lines, for example, then we've done all we need to do." Wrong. You would be just getting started. Thank you.

CHUCK: Thank you, Ambassador Woolsey. You brought forth the whole issue of what local folks can do rather than just wait for somebody in Washington to take care of them. There is something you can do at the local level. I'd like to make certain if there is a question or two before we do the next panel. You could begin to form a line here if there is anyone with a question.

The next panel you're going to see will be including entrepreneurs who are doing various technologies at the local level, some local power generation and storage. You'll see someone doing off grid buildings and also cyber security technology that is just emerging. But I would like if someone is coming down for a question is this is a tough issue, there are a lot of technologies there, but from the standpoint of leadership and what people can do at the local level other than just starting to learn more about this technology that they can use themselves, as we move forward down this path is there anything that you would give us some guidance on in terms of leadership or what to do in the face of other people who are saying, "It can't possibly be a problem because I haven't heard of this before"? What kind of guidance might you give us sort of as a coach in the middle of halftime?

WOOLSEY: Well, I think the key thing is that with electricity to a first approximation there is no upper level. There is nobody in charge. There is FERC that deals with transmission regulation, financial regulation, and has I think an able staff, able people, but the standards such as they are for energy security are kind of sort of set by NERC, NERC, of course, being a perfectly reasonable trade association of utilities. Try to think of the last war you think was won in which the General of one side was a trade association of hundreds of businesses. That's not a good instrumentality for figuring out how to deal with an enemy that may be trying to destroy you. It's not just that it's not good; it's nuts.

There is effectively, there are several people at DoE who deal with electricity, but there is essentially nobody in charge at the national level. You've got 50 public utility commissions, mainly retired utility executives, very sympathetic to the idea that you never want to add any cost, because the public won't like that. As a result of that there is more spent every year on research and development of dog food than research and development of improving the electric grid.

CHUCK: I guess we don't have an appetite for it.

WOOLSEY: Right, yeah. There is all sorts of other associations with dogs one could make in these circumstances. But the point is that if anything useful happens it's almost certainly going to be at the local level. It's almost certainly going to be people beginning to build some resilience into their homes and systems with the solar and batteries in the basement and the rest. And the big enemy of this is the assumption that you've got to get everything, that it's got to be perfect, that it's got to be a complete substitute for all the electricity you could ever want on a good day. And it's the wrong way to think of it. The way to think of it is, like I mentioned in the remarks is how can you keep going.

Now we probably don't want to go this small, but I understand now in Bangladesh it is possible on markets in the street corners and so forth to buy about a $25 dollar kit and that kit contains two or three solar panels, several watts, not hundreds of watts, not kilowatts, watts, a cell phone charger, three LEDs, a tiny fan, and a radio. And what is interesting about that is if you can have LEDs the children can read at night and peruse the school system, the fan can keep you cool, even in a Bangladesh summer if you're in the shade. All of these things are carefully chosen and for an extra about $40 dollars you can get a tiny refrigerator which is largely in those cultures used for refrigerating the medicine for animals, because if you have a refrigerator you'll have medicine and your animals will be healthier than other people's animals and that's a very big deal. Now that is bare bones, okay.

We don't have to go that bare bones in order to have something that is modest but useful and can still keep you functioning and in connection with the world and not freezing to death and so on with a few tens of dollars or hundreds of dollars of systems, many of which might be designed to deal with campers and so forth. There are a number of possibilities here to look at ways to get the very, very basics of what one needs. Probably most people are going to want to be interested in being above that level, but you don't have to have it perfect. And I would put more confidence in average Americans and in the free enterprise system to come up with some of these types of systems than I would anything regulatory having to do with electricity in which so far the more I study it the less confidence I have in the way we do it.

CHUCK: Thank you, Ambassador Woolsey. And I would invite everyone to take a 10 second stretch and give our distinguished speaker a round of applause.

Multi-Flex Fuel Radical-Assisted Engines for Critical Infrastructure Applications with Local Fuels

Bill McCowan, David A. Blank, Thomas Shaw (HCRITI)

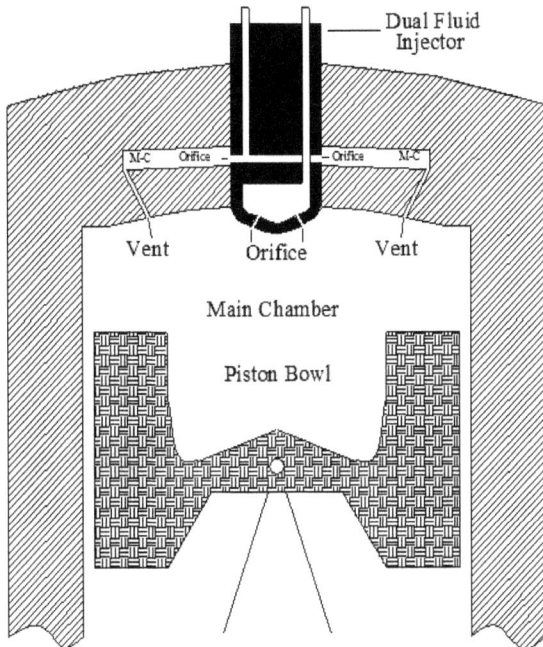

Radical ignition engines utilize chemical processes to control and enhance fuel lean combustion by storing chemicals, called radical ignition species, between the engine combustion cycles. As a result, the fuel is burned more completely and at lower peak temperatures than in conventional engines, making possible 75% or greater reduction in NO_x emissions, as well as improvements in fuel economy. In addition, the technology enables easier ignitability of all hydrocarbon fuels and allows for much greater combustion stability during engine operations with a great variety of fuels. From experience with heavy fuels, gasoline, ethanol (and other alcohols), natural gas, biogas, hydrogen, motor oil, carcinogenic wastes, etc., it is known that this technology can enable highly stable and low-pollution combustion in the same multi-flex fuel engine over a very wide range of such fuels, with very little cycle to cycle variation.

This patented process has been extensively tested via multiple high-end computer simulations, as well as by prototype testing at Argonne National Lab, etc. Using typical locally available fuels, Homogeneous Combustion Radical Ignition Technologies International (HCRITI), LLC, is currently conducting an exhaustive second/commercialization test round with multiple engine configurations suitable to a variety of applications: from stationary engines for use with generators, pumping, etc., to mobility engines for land, marine, and aviation transportation. For power generation, we specifically expect to have a commercialize-able biogas engine within a year.

The implications of radical-controlled engines for the security of critical infrastructure are extensive. Radical-controlled generators will provide a grid-independent power production source that can be run on a variety of fuels, so that if/when conventional fuel supplies are unavailable, locally available/produced fuels such as ethanol, biogas, and hydrogen can be used. These engines will meet the stricter emissions standards planned for the near future while enabling improved fuel economy over conventional generators. With the technology's augmentation retrofit capability, fleets of emergency vehicles can similarly be pre-fitted to continue operation under shortages of conventional fuels. Furthermore, the availability of generators that adhere to strict emissions regulations will encourage the development of small-scale local power generation projects, such as fuel-from-waste projects on farms and in municipal wastewater treatment. Once local self-sustaining power generation becomes widespread, the nation will have much greater capacity to recover from large-scale grid failures.

For further information, please consult the patents and publications listed below, or contact us with questions.

Sample Application Examples:
1. Clean SI-DI Natural Gas HCRI Engine in Brief

In our pioneering Spark Ignition Direct Injection HCRI process, rather than combustion initiation primarily via oxidation, as the flame front progresses, natural gas fuel ignition is instead initiated via "OH decomposition". This process can occur under lean fuel conditions, resulting in much lower local and peak cycle temperatures than typical of the spark ignition (SI) processes of conventional engines. Thus, combustion is more efficient and cleaner. For creating this new ignition process, mini-chambers (M-Cs) are positioned in the head of the engine cylinder so that they are in constant communications with the main chamber. In these M-Cs controlled quantities of radical ignition (RI) species are generated in one cycle for use in the main chamber combustion event of the next cycle.

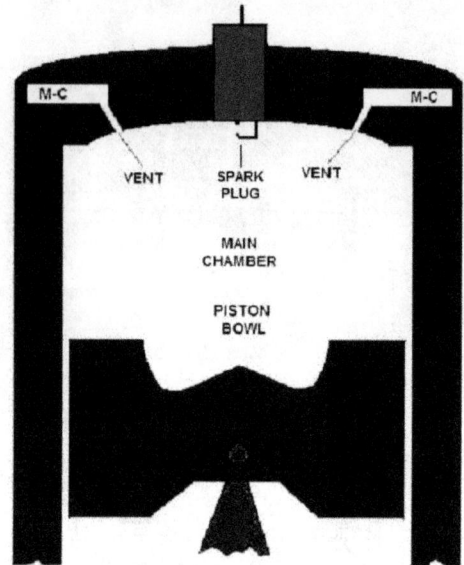

4-Stroke SI-DI Natural Gas HCRI Engine

Short list of advantages:
1. Fuel efficiency improvements (35%–45% over current SI natural gas engines).
2. Near NO_x (nitrogen oxide) free operations with simultaneous CO emission reduction.
3. Combustion optimized for each point within the operating regime of the engine.
4. Combustion leaner and more stable, with less cycle to cycle variation.
5. Engine made Multi-Flex fuel capable.
6. RI combustion control reduces the control regulation measures normally crucial to optimized combustion.
7. The overall engine is simpler and less expensive to both manufacture and maintain.

2. Gasoline SI-DI HCRI Rotary Combustion Engine (RCE) in Brief

Controlled quantities of radical ignition (RI) species are stored and/or generated in mini-chambers (M-Cs) in one cycle of the Homogeneous Combustion Radical Ignition (HCRI) engine for use in the main combustion event of the next (following) cycle. This enables a number of striking advantages:

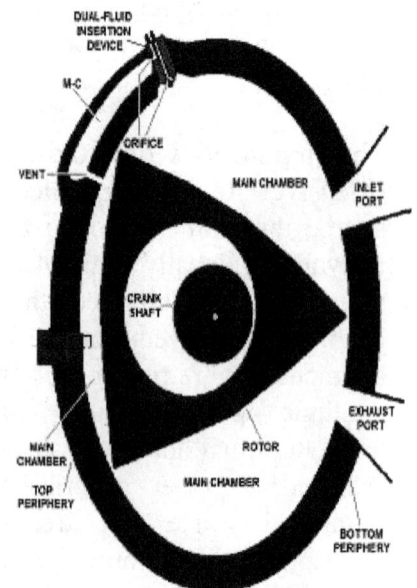

- Fuel efficiency improvements of 25%–30% (compared to conventional 4-stroke spark ignited gasoline engines).
- Nearly soot/PM (particulate matter) free and NO_x (nitrogen oxide) free operations.
- Combustion leaner and more stable, with less cycle to cycle variation.
- Combustion optimized for each point in the operating regime of the engine.

SI-DI HCRI RCE

- RI combustion control reduces control measures normally crucial to optimized combustion.
- The overall engine is simpler and less expensive to both manufacture and maintain.

Instead of combustion in the flame front initiation by oxidation, gasoline combustion is initiated via an "OH decomposition" process.

This process can occur under lean fuel conditions, resulting in much lower peak cycle temperatures than are typical in the spark ignition (SI) and compression (CI) processes of conventional engines.

For this new ignition process, mini-chambers are positioned in the periphery of the rotary engine cylinder so that they are in constant communications with the main chamber.

Patents:
U.S. Patent 7,493,886 B2: "COMBUSTION CONTROL VIA HOMOGENEOUS COMBUSTION RADICAL IGNITION (**HCRI**) OR PARTIAL **HCRI** IN CYCLIC IC ENGINES", February 24, 2009.

U.S. Patent 7,832,372 B2: "COMBUSTION CONTROL VIA HOMOGENEOUS COMBUSTION RADICAL IGNITION (**HCRI**) OR PARTIAL **HCRI** IN CYCLIC IC ENGINES", November 16, 2010.

Most Recent Publications:
D.A. Blank, "NOx Reduction in Natural Gas RI Species Augmented Large Bore Four-Stroke SI Engines", **SAE Paper 2011-26-**0005**, 2011.**

D.A. Blank, "A Novel Two-Stroke SI Design for NOx Reduction in Natural Gas RI Species Enhanced Engine", **SAE Paper 2011-01-**2025 **and JSAE Paper** 20119326**, 2011.**

D.A. Blank, "Combustion Control Chemical-Kinetics Studies with Natural Gas in HCRI Enhanced Four-Stroke DI SI", **SAE Paper 2011-01-**1895 **and JSAE Paper** 20119330**, 2011.**

Biogas from Renewable and Municipal Waste Streams: Applications to Critical Infrastructure

Charles Manto (IAN), David Blank, Bill McCowan, and Thomas Shaw (HCRITI)

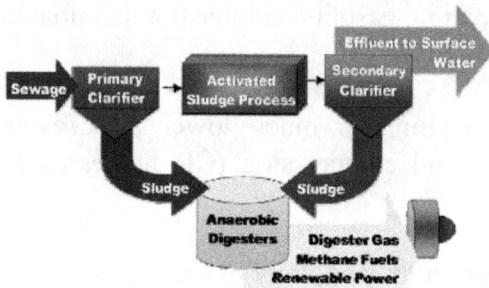

Using waste biomass as an energy source is more than a century old process with large success on an international scale. However, the United States has been slow to develop this technology. For example, wastewater treatment facilities consume about 4% of the electricity in the United States,[1] while many such facilities in Europe generate a significant fraction of the energy required to power them from the wastewater itself. It has been estimated that more complete energy harvesting could yield up to 9.3 times[2] as much energy as our modern plants are consuming. There are also several other biomass waste streams, from food industry waste to landfills, that currently cost energy to dispose of, but could instead be net energy producers.

Instant Access Networks, LLC (IAN) and Homogeneous Combustion Radical Ignition Technologies International, LLC (HCRITI) have a use sharing arrangement in a patent pending process that significantly improves on the anaerobic digestion technology. In anaerobic digesters, bacteria break the large organic molecules in the waste down into methane. The resulting biogas can then be burned in engines to power generators for electric power supply. A series of co-inventions by IAN and HCRITI improve biogas yield by 30% or more compared to conventional digestion processes. This technology thus has the potential to reduce wastewater treatment grid-dependence and/or to act as a power source for any number of desired applications, while at times being able to also sell energy back to the power grid. Our anaerobic digesters are combined with a variety of other technologies, such as an emerging breakthrough engine technology, smart micro-grid controllers, and wind plus solar energy augmentations to provide reliable and cost-effective electricity. We are also currently working on an application that can leverage even greater than previously anticipated[2] energy yields from available biomass sources. Prototype digesters are already in operation at several sites, and several further test sites are planned to open in the next year.

Furthermore, energy recovery from biomass has far-reaching environmental benefits. For example, it reduces methane emissions and other pollutants that may be released into the environment from waste streams. One molecule of methane has 21 times the global warming impact of one molecule of carbon dioxide. Also, converting waste to energy alone has the potential to accomplish the U.S. federal administration goal of 25% renewable energy by 2025.[1]

Perhaps more importantly, our biogas technology has widespread implications for critical infrastructure. It is important that critical infrastructure have dependable local power supply to reduce grid dependence and to enable uninterrupted operations during grid failures. While various options for local power generation exist, many of them are expensive and require extensive infrastructure to produce enough electricity. Anaerobic digestion represents an excellent solution for providing energy to critical infrastructure applications: as it can cost effectively produce enough energy to supply needs from sources that are both renewable and locally available.

Sample Application Examples:

1. Novel Municipal Waste Power Production System

Towns, Cities, Prisons, Military Bases, Hospitals, and College Campuses are among examples of human conglomerations that produce substantial biomass waste with high energy content. The proprietary patent pending system evolved by IAN and HCRITI for such applications has a number of benefits and energy conversion advantages.

Short list of advantages:

1. Convert pollutants to renewable products via significantly lower energy input.
2. Reduce cost of wastewater treatment.
3. Dramatically increase the amount of methane-derived energy from wastewater.
4. Generate 30% more electricity from municipal waste water than current conventional systems.
5. Make sewage/waste water treatment a net power producer that is thereby independent of the power grid.
6. Reduction in greenhouse gas emissions from waste water sources by more than 80%.
7. Enable hybrid wastewater treatment strategies that incorporate other renewable energy technologies within bio-digestion.

2. Renewable Biomass Power Production System

Some enterprises, such as the food industry, have inherent biomass streams, often linked to large energy demands. The proprietary patent pending system evolved by IAN and HCRITI supplants the necessity for complex waste processing equipment, while supplying energy to the enterprises in question.

Short list of advantages:

- Converts organic matter into renewable energy at higher yield ratios than conventional.
- Enables inexpensive renewable energy and water storage for contingencies.
- Generate 50% more electricity from biomass than current conventional systems.
- Enable local enterprises to operate independently of the power grid.
- Reduction in greenhouse gas emissions from select local enterprise by more than 85% offset costs to enterprises for biomass processing by producing energy.
- Enable more holistic renewable energy solutions.

References:
[1]NIST, "Energy: Technologies to Increase Efficiency of Wastewater Treatment", 2010
[2]WERF, "Energy Opportunities in Wastewater and Biosolids", 2009

Grid-Independent Buildings and Communities
John Spears, President, Sustainable Design Group, International Center for Sustainable Development

The Problem

We live in an unsustainable world. Since the mid-nineteenth century, the developments in technology, the growth of the world's economy, and the quality of life we enjoy today have developed almost entirely around the consumption of fossil fuel. In order to see why this is unsustainable, we can characterize man and society in the twentieth century as one big system or a machine. The machine has energy inputs to the system that creates the work or benefits of the system. At the other end is the output of the waste or leftovers from the process—the exhaust pipe. This is a linear system. Energy in–Work or Benefits of the system–waste out.

In the modern world, the energy into the system is primarily fossil fuel we have extracted from the earth. Fossil fuel is a limited resource. We only have so much. Many argue about how much we have left, but it is an undisputable fact that it will eventually run out. Then what? Basing our entire society's well being and quality of life or even survivability on a supply of fuel that will eventually run out is short sighted and irresponsible considering that viable alternatives, renewable sources of energy, exist today such as solar, wind, and biofuels.

At the other end of the system is the waste from fossil fuels. Burning fossil fuel creates deadly toxins that pollute the air we breathe and destroys the ozone layer of the planet that keeps the earth's climate suitable for life to exist. The drilling, mining, and handling of fossil fuel pollute the water we need to drink and destroy the soil we need to grow our food. In short, the fuel that powers the society today is running out quickly and is poisoning us and the planet in the process.

The Solution

The life style and economic growth we enjoy today can be fully powered by renewable energy sources. Numerous studies have shown that a 100% renewable energy economy is possible with today's technology and is an excellent investment for economic development. One example at a country level is the study called "Energy Rich Japan"
http://www.energyrichjapan.info/pdf/EnergyRichJapan_summary.pdf
"Using baseline data from 1999, the 'Energy Rich Japan' report shows how a combination of the best energy efficiency technologies available today, and a massive investment in renewable energy, could ultimately provide Japan with 100% of its energy needs from renewables—including transportation fuels—without expensive and environmentally damaging imported fossil and nuclear fuels. Rather than seeking energy security through its hugely expensive and polluting nuclear program, for example, Japan could instead build its own renewable energy industry. As an energy-hungry and supposedly 'resource poor' country, Japan could make this transition to clean, renewable energy without any sacrifice in living standards or industrial capacity."

Another study at a State level, done by the author for the State of Maryland, looked at the economic development potential of clean energy for Maryland:
http://www.solarcities.org/pdf/MCECStudyReport2-28-07.pdf

This is the most comprehensive study of energy and the potential of renewable energy ever done for the State. The report concludes that Maryland could produce from 30% to 137% of all the State's electricity from solar photovoltaics (PV) and on-shore and off-shore wind power at costs often competitive with conventional sources.

At a community level, we have shown that sustainable communities can be developed based on 100% renewable energy and with strong economic development opportunities. One example is the Model Sustainable Community project done by the author for the Asian Pacific Economic Cooperation (APEC): http://www.solarcities.org/icsddoc/longjumasterplan.pdf
This project demonstrated to the Chinese the principals of sustainable community development from both an energy and economic development perspective.

At a single building level, there are numerous examples of energy independent, zero energy buildings. The Earth Home http://www.solarcities.org/earthhome.html is an example of a completely self-sufficient home that provides the occupants with electricity, clean water, and sanitation. We have built thousands of these homes in the developing world as well as homes in the Washington DC area including the author's home and office in Gaithersburg, Maryland. http://www.sustainabledesign.com/manor-house.html

We have just completed the first of its kind, completely off grid commercial building for Frostburg State University. http://www.fsuwise.org/renewable/SERF/ This 6400 square foot building is powered by wind and solar energies and has batteries and a fuel cell for backup power. The Sustainable Energy Research Facility (SERF) building is completely independent from the grid and requires no fossil fuels.

In Frederick, Maryland, we are building 58, affordable zero energy town homes that are 100% solar powered. http://www.sustainabledesign.com/nexus.html

What can we do now?
The electric grid infrastructure is aging and increasingly vulnerable to failure due to severe weather and equipment failure. The electric grid is also very vulnerable to attack by terrorists. Our buildings are normally 100% dependant on the electric grid for heat, light, communications, refrigeration, water, etc., basically everything. When the electric grid goes down, everything basically stops working. We have seen this scenario all too often and it seems that the black outs are becoming more frequent. What can be done now?

Homes, offices, businesses, and community services need to begin to transition to having the capability to become grid independent. Grid independence is the ability to provide all the basic services needed indefinitely when the grid goes down. Homes need to be livable, businesses need to continue to operate, and community services need to be uninterrupted. When the grid is available, the grid-independent building draws power from the grid and in many cases supplies power back to the grid when its solar and/or wind system is generating more than it needs at the time. When the grid goes down, all basic functions of the building are provided by renewable energy sources and batteries. Gas generators are not grid independent. As we have seen in all major utility disruptions, fuel for generators quickly runs out due to high demand and lack of power to run the pumps to pump the fuel.

The design of the grid-independent building starts with a very energy-efficient building to reduce the heating and cooling load. All the appliances and lighting are selected to be of the highest efficiency. A high-efficiency building requires much less power and therefore requires less solar or other renewable energy.

Heating can be provided from a number of sources. The first is passive solar heating. Passive solar heating is basically designing the building so that the sun coming through the windows will provide most of the heat and light during the day. The building is constructed with masonry to store the heat for the evening hours. Passive solar heating can provide 50%–80% of the heating load. Back up heating can be provided by wood heat. Wood is a renewable resource and carbon neutral. Heating and cooling can also be provided by a very efficient geothermal heat pump powered by solar PV.

Hot water can be provided by a solar water heater and also from the geothermal heat pump. Electricity for lighting and appliances is provided by a solar PV system with a battery backup. In an energy-efficient home, the system can be relatively small. For example, the town homes in Frederick require only 4 kW solar to provide 100% of the electricity including the geothermal heat pump. Cooking with electricity can require a large amount of power and tax a solar/battery system. Microwave cooking and induction ranges are the most efficient. In some cases, it may be practical to have a propane tank to provide backup cooking energy.

Water is a vital resource for the grid-independent home. Central water supplies may be compromised and therefore the grid-independent building should have its own independent water supply. If you have access to a good well, a solar-powered pump can pump the water. In most of our grid-independent homes, we provide a rain water collection and storage system sized to provide 100% of the water requirements for the home.

Sustainable Community Centers

A Sustainable Community Center (SCC) is a place people can go during an emergency if their home is uninhabitable. The SCC is grid independent. The SCC has an independent solar power system with batteries, clean water cistern, and sanitation. Being energy independent, the SCC can provide services to the community during an emergency. SCCs can be schools in the community that are redesigned to be grid independent. During non-emergency times, the solar power system will power the school and reduce the energy costs for the school district. In addition to SCCs, fire stations and police stations should also be grid independent so they can function independent of a failure of the grid.

Economics

We have been demonstrating cost-effective grid-independent homes for years. Using the Frederick zero energy town homes as an example, the grid-independent home is less expensive than a typical home on a monthly cash flow basis. If you add the monthly mortgage payment and the monthly utility bill together, you get the total cost of home ownership. The grid-independent home costs more, and therefore has a higher monthly mortgage payment, but this is offset by the lack of a utility bill. For the Frederick town Homes, the solar-powered zero energy homes have a monthly mortgage plus utility cost of about $90 less than a comparable new town home in the area. In addition, the Federal and local incentives for solar and geothermal add up to about $12,000 per home.

Conclusion

We need to eliminate the use of fossil fuels as fast as possible and shift to a renewable energy economy. This is the only sustainable path to a healthy and prosperous society. We can do this today with existing technology in a cost-beneficial way. The technology we are using today for grid-independent homes has been around for 30-40 years. We have demonstrated that solar-powered homes are less expensive than conventional homes.

Secure Portable Systems Devices
Wendy Richards, Chief Business Development Officer, Sky Catcher Solutions

1. Executive Summary

This paper describes the development and integration of a technology originally developed for the Department of Defense to allow users and organizations to deploy a highly secure, mobile virtual computing platform to be used with mobile hardware or in an Enterprise environment. Throughout this paper this technology and its' embedded solutions will be referred to as a *Portable Systems Device*™ (PSD).

The applicability of this technology to the end-user, and ultimately to an entire Enterprise, is multifaceted. Enterprise organizations all over the world have long asked for a solution that will meet not only their Data-at-rest (DAR) requirements, but also provide for a computing platform that is secure and enterprise manageable. This virtualized computing platform resists data contamination from the host machine inside a protected enclave, at a residence of a teleworker, or in the classroom. Another preferred convenience for a hardened solution of this nature is that it be small and ultra-portable to facilitate the movement of the device from one location to another without the need for re-registration or reconfiguration. Data-in-transit has always been a key area of concern. Data on sensitive networks should be protected while it is in motion and data should be able to traverse the Internet when personnel are not directly connected to or through their Enterprise.

The core of our approach was to create a product unlike any other available on the open marketplace. Our technology creates a virtual computing platform that provides extremely strong authentication, is FIPS 140-2 certified, and is secured by two layers of hardware encryption; AES-256 bit and Anti-Statistical Block Encryption (ASBE) 2008 to 2 GB. The PSD solution meets stringent Department of Defense standards, is highly portable, and leaves zero residual footprint on the physical host machine. The solution is also enterprise manageable not only to the PSD platform, but also to security controls put into place by Security Administrators within the Enterprise.

This technology is ready for any environment and deployment: This paper defines the capabilities created by this technology. It will also define different options to implement this solution and discuss the strengths and weaknesses of those options.

The technologies described in this overview will be as follows:
> ➢ Secure Portable Systems Device Types
> ➢ Two layers of encryption technologies in place
> ➢ Enterprise Management Capabilities
> ➢ Alternative uses as a secure storage device ONLY (if desired)

The paper will further discuss this PSD implementation including:
> ➢ Section 2: Solution Types, Scope, and Options Available
> ➢ Section 3: PSD Configuration Types
> ➢ Section 4: Use Case Studies
> ➢ Section 5: Conclusion

2. Conceptual Ideas and Components

The technical approach to virtualizing and securing these computing platforms is multifaceted. It had to be user-friendly and intuitive while at the same time maintaining ALL security requirements. All technologies bring with them new challenges not only to the end user but also to the System Integrator (SI) as well. The first technical hurdle to overcome was to offer an interface that was not only familiar to the user, but also secure from the onset. The platform needed to fit into the security framework that has already been established by the Enterprise, manageable by current Systems Administrators (SA). Introduction of this virtual computing platform will present a familiar environment to all users and Systems Administrators, allowing them to immediately benefit from its integration. There will be no learning curve or ramp up time on its use. The computing platform can, however, be tailored to almost any Operating System (OS) and to almost any configuration. One of the most important aspects of the computing platform is that it must meet all customer, security, and operational requirements. When the solution is implemented, it should not affect current implemented programs or user experience.

The second technical hurdle was to adhere to Information Assurance (IA) specifications while maintaining usability and familiarity to the end user and administrators, and to alleviate the need for training of new systems or components. Following requirements from the Department of Defense and Federal guidelines as well as other IA requisites, this Portable System Device solution is fully security compliant to any scenario it is introduced into, including the Department of Defense. This solution, however adopted, is also secure when it is not in use. The data on this virtual platform are secured by an onboard hardware encryption processor which provides two layers of encryption. The first layer of encryption is a strong AES-256 bit encryption, which is already certified to the FIPS 140-2 standard. The second layer is the next-generation encryption, Anti-Statistical Block Encryption (ASBE), which fundamentally future-proofs the virtual computing platform. If a device is lost or stolen, there is no chance of the data inside being compromised. Aside from the unsurpassed strength of the dual encryption, the solution has enterprise management capabilities that allow both Security and Systems Administrators to perform their jobs effectively from solution initiation without the need for additional training. On a network, the device functions just as any other host computer would act on a network. Maintenance of the devices themselves is accomplished from the device enterprise management console. Device enterprise management offers role-based administration and the ability to perform a variety of functions, from resetting a password to remotely wiping a device clean if it is lost or stolen. This offering is a computing platform that is secure, IA compliant, and enterprise manageable.

In addition to the solution being FIPS compliant with strong encryption and the data secured onboard the device, authentication into the device must be just as effective and easy to accomplish. The design of the device is such that each device is capable of more than 10-factor authentication. Devices that have biometric capabilities embedded in them coupled with smart card (CAC/PIV) and password authentication provide a high degree of assurance and security, with the ASBE adding an additional >10-factor authentication capability. This truly enforces the "We Know Who You Are" rule of authentication. Technical hurdle number three is exceedingly well addressed ensuring that the person accessing the device is in fact an authorized user.

The fourth technical hurdle that had to be overcome was the challenge of securing the communication channel, even when a user is not in his native AOR (IE: Not directly connected to the Enterprise network or when being used by a teleworker or student). Since protection of data-in-transit is a major security concern within any organization, a unique VPN tunnel was developed. It will conceal and secure the transmission of data within that tunnel in a FIPS-approved security architecture. Since the client is based on the OpenVPN platform it is highly customizable and can be tailored to the unique situation within the architecture it is placed. The client can be integrated into the PSD as part of the default configuration with organizational security certificates included. Additionally, a second option for creating an invisible and impenetrable wired or wireless network is available for those end-users with the need for a completely secure and private network environment.

2.1. Scope
The scope of this solution is to:
- Present an ultra-mobile "go anywhere" platform that can travel with a user into any environment in which they are placed.
- Secure this platform with FIPS certified encryption modules.
- Have in excess of 10 factor authentication if desired.
- Provide a secure communication tunnel no matter where the user is connected and performs device anonymization, or connect to a completely secure and private wired or wireless network.
- Maintain all IA standard and remain IA compliant as delivered to the Enterprise, and maintain that compliance throughout its lifecycle.
- Facilitate the beginning of a new standard in which a virtual computing device is issued as part of a basic employment process or utilized as an alternative to extremely vulnerable home computers. The PSD is capable of holding authentication credentials, network access controls, e-textbooks, biomedical data, and much more while staying with a user as long as they carry the device.

2.2. Conceptual Architecture
The baseline for this concept is simple: computing platforms that will facilitate thin client type access and allow secure portability and protected communications. There are basically two different scenarios regarding how this solution could be deployed. Both have similar enterprise management; offer greatly enhanced encryption and the ability to protect data-in-transit. Both will be discussed in the following paragraphs.

2.2.1. Microsoft Windows Embedded Architecture
This architecture utilizes a portable USB hardware platform that boots from a Microsoft Windows pre-boot environment on a host computer into a pre-boot authentication application. After successful authentication, the device reboots the host computer and arrives in a native Microsoft Windows Embedded Operating System. This embedded environment is almost identical in nature to Microsoft Windows XP Professional. This scenario solves the familiarity problem by presenting the user with a familiar interface, and it offers driver support for over 12,000 devices. The Operating System, if configured correctly, will not allow access to the underlying host hardware. This virtually eliminates the chance of

contamination from the underlying host computer. In addition, Windows Embedded (WES) devices offer enhanced write filters to help prevent users from becoming infected by Internet-borne threats or from installation of unauthorized software of any kind. The Operating System can have multiple standard configurations, keeping in mind that hardware support may be limited due to the need for separation of the physical host components. In other words, if a driver is loaded that offers access to the physical host hard drive, there is a chance of potential contamination from that drive, if a driver to the physical host is not loaded, then there is very little chance. In all, hardware configuration on a WES device can be extremely difficult to configure. Software can be loaded and run as in a normal computer and documents can be stored on the device itself or shared across a network connection to a cloud. Virtual devices such as CD-ROM libraries are difficult to load and maintain because of the lack of support for virtual device drivers. If new software or security updates are required, the device is launched in a "maintenance" mode which allows new software updates to persist across reboots. If an OS becomes corrupted, the device can be reimaged with its base image, again providing for minimal downtime.

Administratively, the WES solution provides for the same administrative functions offered in a physical host running Windows XP. From a network management standpoint, there is no difference between the two. The only real role for WES is teleworkers. It is not readily capable of a variety of other roles such as a forensic platform or a first responder device.

2.2.2. Linux Boot Running Virtual Operating System

The other type of configuration available is a Linux booting Operating System. In this configuration, a PSD boots from a Linux (usually Ubuntu) pre-boot environment on a host computer into an authentication application. After successful authentication, the device launches VMWare player and then a virtualized host operating system. That Operating System can be any OS from Windows XP to Linux. The only consideration would be the minimum required hardware to run an OS. For example, running Windows Server 2008 from the device would be ineffective unless the physical machine were capable of running Windows Server 2008. Generally speaking, Windows XP and most Linux variants are safe to run on almost any hardware found in any environment. Since driver and kernel support is provided by the underlying Linux OS, nearly all of the hardware available on the open market is accessible to the Virtual OS.

Security is enhanced greatly by running a Linux Boot device as well. Not only is the base Linux image secured to DOD standards, but the Virtualized OS can be tightly configured and secured to DOD IA directives as well. Hardware can be managed by the underlying Linux OS AND the VMWare image to restrict access to any or all of the physical host machine. This means the solution can take on multiple roles such as a forensics platform or first responder device. If image corruption occurs, the feature of VMWare to "step back in time" to an earlier snapshot of the OS will significantly minimize downtime. Software can be run and loaded by users, and documents can be saved to the device or to the cloud. This can be controlled by Information Assurance management components. Virtual devices can be loaded and maintained easily. In addition, software write filters can be engaged in a VMWare host to disallow writing to critical system files.

3. PSD Configuration Types

Since there are multiple configurations, cost savings can vary from a slight savings on current technology to a significant cost savings depending on which platform is chosen. It is notable that a Windows Embedded solution has already gained Department of Defense acceptance. A Linux booting solution running a Microsoft Windows Virtual Machine is nearly identical in security function, but generally more customizable and offers a better security platform to work from. The following section will analyze the benefits and weaknesses of the two basic kinds of configuration for a PSD.

The following capabilities/feature matrix shows the comparison of features. Immediately following the matrix, each of the features will be explained in detail.

Feature	Microsoft Windows Embedded	Linux Boot Running Virtual Operating System
Pre-boot Authentication	Y	Y
Customizable Virtual OS	P,D	Y,E
Robust Security	P,E	Y,M
Hardware Customization	P,D	Y,E
Security Administration	P,D	Y,E
Enterprise Manageable	Y,M	Y,M
Device Role Customization	N	Y
Choice of Operating System	N	Y
Additional Cost	Y	P
Tamper Resistance	P	Y
LEGEND	**Y** = feature fully supported; **N** = Feature not upported; **P** = feature partially supported; **D**=difficult to support; **M** = medium difficulty to support; **E** = easy to support	

3.1. Feature Matrix Breakdown
3.1.1. Pre-boot Authentication

In either scenario described pre-boot authentication is easily accomplished. The pre-boot environment launches an authentication program that users will authenticate to, based on the level set by the Systems Administrator. This authentication can be in excess of 10 factor authentication. Upon successful authentication, the pre-boot environment will execute the user boot environment and the user will be placed into the OS running on the device.

3.1.2. Customizable Virtual OS

Although the Windows Embedded Operating System can be customized and a "golden image" created, the customization becomes difficult at this point. This is because of the write filters active in a WES device. Since the write filters do not allow ANY system areas, new software loads from the users are not possible. Administrators would, in most cases, have to physically take the device if additional software loads were needed. Administrators would then have to place the device in a "maintenance mode" to load that software. During this time, the user is without their device and productivity is lost. With a Linux pre-boot environment, a golden image is created much like the WES devices, it is mass duplicated to the devices, and the devices are issued to the users. Users can load software based on their administrative rights on the network. This more closely matches how a typical physical computer would work. If write filters were desired, they could be implemented much more effectively and targeted to just system folders or files.

3.1.3. Robust Security

WES devices offers an additional single layer of protection. The write filters that are enabled prevent users from writing to or changing system files or folders. This is effective and offers adequate protection from malicious insiders. With a Linux device and a Virtual Host, administrators are offered more choices and a greater protection profile. This scenario almost completely prevents even a skilled, but malicious insider from tampering with the device and making unauthorized changes.

3.1.4. Hardware Customization

Windows Embedded offers device support for over 12,000 devices. This offers a great deal of hardware support. However, when the intent of the device being used is such that there will be no chance of cross contamination from the host computer, the idea of hardware customization is severely hampered. Loading device drivers to support physical host CD-ROM opens the computer up to the potential of physical host cross contamination from the physical host hard drive. The chance of contamination is relatively low, but it does exist. Because of the fact that a Linux solution has a virtual guest OS running on it, that guest can be configured to disallow the discovery of ANY other hardware, other than the desired hardware. For example, if the guest OS is configured to see only the physical host CD-ROM device, that is all it will see. There is no chance of opening up other devices either by mistake or by malicious intent.

3.1.5. Security Administration

Security is nearly identical on each of the two platforms; however, application of security updates and patches is quite different. When using a Windows Embedded OS, the devices must be placed into maintenance mode. This requires Administrator interaction and is not easily accomplished from a central management console. This is due to the fact that the devices must be shut down first and then put into maintenance mode BEFORE application of the patches. With a virtual guest OS, the patches are applied exactly as they are today and the Enterprise programs that manage them can be used to do that task.

3.1.6. Enterprise Manageable

Both solutions offer an enterprise management capability. Both have management capabilities that offer the ability to remotely administer the device. Passwords can be reset if forgotten, devices can be remotely administered, and user maintenance can be performed.

3.1.7. Device Role Customization

The WES devices are best suited for teleworker application. They offer little capability other than that role because of the inability to interact with the underlying kernel. However, Linux boot devices offer almost countless possibilities when choosing what role the device will perform. Linux devices can be configured as biomedical devices for hospital personnel and secure patient data; they can be configured as First Responder devices for Fire and Police personnel and, of course, for teleworker application. They can be integrated into SCADA systems, act as secure video surveillance platforms, or even as libraries for mobile training teams. As a platform, the Linux Boot solution is the "Swiss Army Knife" of devices. It can assume almost any role it is placed into.

3.1.8. Choice of Operating System

WES devices boot natively into a Windows Pre-boot (WinPE) environment and, after authentication, boot into Windows Embedded Standard. That is the only option available when utilizing a WES device. Conversely, Linux boot with virtual guests offer the ability to choose almost any Operating System to boot into. As long as the physical host the device is placed into can accommodate the device duty, the OS will work.

3.1.9. Additional Cost

WES devices are not covered under the current volume license agreement and therefore an additional Windows license must be purchased for each PSD purchased. This represents an additional expense that cannot be mitigated. That expense could prove extremely costly and is not necessarily warranted. With a Linux Boot solution, the underlying Linux OS is free and covered under GNU public license. The virtual guest could be used from the existing Volume License Agreement with Microsoft and would not represent any additional expense. The only time any additional cost would be incurred would be if there was a license cost for a particular program and that license was not already purchased by the customer.

3.1.10. Tamper Resistance

Both devices are physically tamper resistant. They both are hardware encrypted and physical excavation of the devices will render them useless. Tamper resistance of Linux boot devices is a bit more robust when it comes to insider threat. When using Linux devices, a user will not be able to "break out" of the security constraints they are placed within. With a Windows Embedded device, a skilled user could potentially cause the device to start searching for new hardware devices and place the security of the device in question by opening it to contamination from another device.

3.1.11. Dual-Layer Encryption

The PSD provides dual-layer encryption. In addition to the standard AES-256 bit encryption, the PSD also provides an additional layer of encryption. Fortified with Anti-Statistical Block Encryption (ASBE), this next generation solution is not subject to attack models and methods of

Cryptanalysis. Standard statistical analysis and any attempt at byte frequency cannot crack this advanced encryption. ASBE protects data with variable-length keys that scale between 2008 bits and 2 GB, and passwords up to 64 kB. Keys and passwords are 'generated–destroyed–recreated' on demand, making transfer between end points unnecessary and circumventing criminal detection or interception

4. Use Case Studies
In this section, we demonstrate how the solution might operate under different operating conditions within an environment and integrate with current Information Assurance (IA) Computer Network Defense (CND) and forensic tools that can be leveraged to enhance its capabilities.

Use Case Study 1: A law enforcement agent is assigned to investigate a possible computer crime. That agent normally has to gather his arsenal of tools when he travels. He needs a laptop/touch pad, his media library, investigative tools, etc. With the PSD solution, there is no media library or other software tools to gather. It is all contained on the PSD, securely stored in a partition that only the agent can access. Depending on the device used, the authentication to this partition is ensured by potentially more than 10 factors of authentication. The agent simply collects his laptop/touch pad and his PSD and goes. All the tools needed are with the agent and he can begin immediately after arriving at his destination.

Use Case Study 2: Rather than purchase traditional textbooks, a college student decides to go with e-textbooks instead. The student needs to access sensitive information across the Internet and a wants a secured platform to access it from. Even though he may have been issued a laptop, that laptop is not secured and any information stored on it is vulnerable if it is lost or stolen. With the PSD solution, the user can have a completely sterile laptop and still have sensitive information with him at all times. The laptop merely becomes the presentation layer for the "real" computer located within the PSD. The virtual machine on the key can be configured to fully mirror the capabilities of ANY computer the user would normally carry. Also, because of the strong multi-factor authentication capabilities coupled with the onboard hardware encryption, the data is SECURE even if the PSD is lost or stolen. If such an event were to occur, the back-up of the last session can be recovered from the enterprise cloud quickly and easily. The virtual machine can carry ALL the associated files and media that the user would normally carry AND provide stronger security measures.

Use Case Study 3: A member of a Navy SEAL Team is deployed to a hostile area where his communications are subject to monitoring and he runs the risk of sensitive information being compromised, or at the very least someone "eavesdropping" on his communications. The Navy SEAL, using his hardware encrypted virtual machine, launches a unique, secure virtual private network solution contained within the virtual environment on his PSD. His communications are transmitted via a network tunnel encrypted with both AES-256 and ASBE that conceals his identity. That encrypted communication ensures no one can intercept his transmissions or eavesdrop. In addition, he could securely access a highly secure private network limited only to designated users.

Use Case Study 4: The Washington, DC area is shut down due to a snow storm. Personnel assigned to this area are told that facilities will be closed indefinitely until it is safe to travel.

Normally this means that there is countless lost workdays as personnel sit at home and wait to return to their duty station. If personnel were issued portable systems devices as a remote worker solution, they could simply insert the PSD into their home computer or laptop and be instantly and securely connected back to the corporate or Government environment without the risk of contamination from their personal computer. Furthermore, this scenario applies to a pandemic event as well. PSDs issued to employees would provide them with secure telecommuting capabilities from any host computer, providing a pandemic solution for corporations and agencies.

Use Case Study 5: Within a regional medical facility or a University Health System, there is the need to protect sensitive data. Data breaches like those seen recently at major healthcare providers such as WellPoint, Blue Cross-Blue Shield and at the Veterans Administrations had a serious impact on the identity security with the patients under their care. With a PSD, a patient would have their entire Personal Health Records (PHR) with them at all times. In the case of military personnel, a PSD would negate one of the longest standing vulnerabilities within the VA. Health records or personnel records in a hardware encrypted format on a device means no more carrying around the original paper copies of their records which puts those records at risk for loss, theft, damage, or destruction. When checking into a healthcare facility, the patient or veteran simply uses a PSD to update their record status and check into the facility. When the patient action is complete, the device is updated with the new/updated information.

These are but a few of the use cases for the PSD. There are countless more applications for the device. The device can quickly morph to become almost any set of tools or a virtual computing environment.

Section 5—Conclusion

In conclusion, both types of device configuration have a place within corporate, law enforcement, government, military, medical, or even home computing environments. It is the key to telecommuting on a secure environment from any untrusted computer (i.e., a user's home PC) or performing even the most demanding tasks such as forensic acquisition from a trusted, secure, and portable source. In events like the ones that have been seen in the Washington, DC area over the last couple of years, when personnel are not able to reach their assigned work location due to severe weather, devices like the PSD would have proven invaluable.

Now, personnel can work from their home, hotel, coffee shop, or cyber cafe while still working from a secure, trusted platform, even though the physical device or wireless Internet they are working from may be an untrusted host computer or unsecured Internet access, whether close to home or abroad. First responders could use the device as part of their kit, allowing them to carry libraries of information or other sensitive data without the need for physical media that could be lost or destroyed. The scenarios are endless.

The primary benefit is that the devices are virtual. The benefits of virtual computing would be beneficial throughout global markets. Those benefits are:
- Huge cost savings in equipment, electrical, and personnel
- Improved Information Assurance standards
- Enhanced access controls by implementing additional authentication methods

- Can use up to 75% less physical space
- Can use up to 75% less physical equipment
- Up to 40% reduction in manpower
- Rapid deployment capability
- Quickly load, copy, or restore a virtual machine to another device
- Maintain virtual libraries without the need for physical media
- Ability to transform a physical host computer into a virtual computer and mass produce it
- Rapid transition back to full capability from a Disaster Recovery scenario.

The conclusion that is drawn from this research team based on the data examined here and the need to put the right tool within any environment is clear: Windows Embedded (WES) devices are an adequate platform for teleworkers, students, and clerical staff that do little more than surf the web, process documents, and send/receive Email. The WES devices are however highly vulnerable to potential security gaps that, if exploited, could compromise the trusted platform that was created to safeguard data. In addition they do not offer the flexibility to perform multiple roles. The Linux bootable devices running a virtual guest Operating System, offer greater flexibility, stronger security, and better value. Both have their places, but generally speaking the Linux bootable device is a better choice for enterprise deployments on a global scale.

Cyber Security
A. Curt Massey

Impenetrable Network Access… Encrypted Node Security

Closing the vulnerability gaps of traditional network security, *STTealth* Shield™ merges NAC with dynamic encryption protection on every computer node on the network. This pioneering software-only solution safeguards subscribers against espionage, from both inside and outside their network.

Overcoming Five Major Weaknesses in Network Security and Encryption

Weakness #1: *Networks and firewalls provide detectable information hackers can access.* Networks that are not protected with *STTealth* Shield provide hackers with information such as IP address and open ports. When hackers acquire *some* information, they use this information to make further attacks. For example, when the OS and patch level are known, the hacker can determine the security flaws that remain. Any open port is also an open door into the machine! The Internet has become increasingly complex, leaving many enterprises vulnerable to malicious attacks. Every year, network security breaches cost 100's of billions of dollars.

Compromised information has resulted in serious threats to the security of nations. Organizations traditionally have responded to these threats by enhancing network security through the use of Firewalls, Virtual Private Networks (VPN) devices, Anti-Virus applications, and Intrusion Detection Systems. Organizations are continually looking for more dependable, scalable solutions to broaden their reach and increase their protection.

Other systems respond to attack probes with a message denying access, which confirm open ports and other information necessary for a cracker to penetrate the system. In a Denial of Service attack, targeted systems require more and more processor usage to respond to massive, multiple, unauthorized incoming traffic until the system is overloaded and shuts down. *STTealth* Shield is the only known communications security system that simply ignores unauthorized traffic.

The Art of Camouflage: Not a firewall… more than white-listing… *STTealth* Shield does not respond to port probes, effectively ignoring DoS perpetrator attacks, fake IP attempts, and all unauthenticated and unapproved network messages. Clients are concealed from the Internet by an orb of imperceptible security. All activity within *STTealth* Shield's "invisibility cloak" cannot be observed or ascertained by unauthorized users.

Rogue Chips Rendered Useless: Embedding malicious code and "back doors" into micro-chips is a growing trend in espionage. *STTealth* Shield controls outbound traffic, as well as inbound traffic, obstructing all communication between the rogue chip and its criminal host. Any traffic attempt triggers real-time alerts, identifying both the attempted destination and the affected machine.

Weakness #2: *Short, fixed, identifiable keys are easy to crack.* Common encryptions, such as DES, RSA, and AES, produce simple short key strands, which continually repeat in ciphertext. These fixed length keys are detectable and routinely broken.
STTealth Shield uses Anti-Statistical Block Encryption (ASBE) with variable key-length that scales from 2008 bits to 2 GB.

Weakness #3: *Encryption key communication is detectable and predictable.* Other cryptosystems require that keys are sent back and forth between users or computers. *STTealth* Shield key generation, communication, and storage cannot be detected, as keys are generated–destroyed–recreated, on demand. This eliminates a need to communicate the key, circumventing criminal interception.

Weakness #4: *Unencrypted data can be analyzed by protocol definition.* Current VPN technologies provide static encryption, which means the key is generated once and shared throughout the system. This provides a certain level of security, but once that key is broken, attackers can intercept and decipher information freely. Criminals can then configure their attacking systems to use the same encryption, enabling them to have a secure method to execute future attacks.

Anti-Statistical Block Encryption: Relentlessly Transforming Security

Data security in the contemporary business environment is increasingly complex, requiring acute acumen and calculated strategy. With new levels of severe and sophisticated threats penetrating the most fortified systems, a drastic change in security tactics is mandatory.

STTealth Shield meets this demand with an embedded encryption engine that is vastly more powerful encryption, with unpredictable and changing keys, which defeats increasing computer-power and criminal investigative technique. Utilizing Anti-Statistical Block Encryption (ASBE), the algorithm is not subject to attack models and methods of Cryptanalysis, not based on mathematical technique, and not subject to statistical analysis.

No two encryptions are alike: Each encryption process *always* results in a different cyphertext with varying length, even when repeating the same plaintext to encrypt, key, and password. In addition to variable length keys from 2008 bits to 2 GB, ASBE allows scalable passwords to 64 KB. The ASBE Encryption Platform is an under-the-hood powerhouse for agility and speed. The cryptosystem platform's data generator outputs keys and passwords that are "generated–destroyed–recreated" on demand, eliminating transfer between end points. The platform incorporates dynamic multifactor authentication and is run by a scripted controller that "wraps" the process into a tightly customized and impenetrable sequence of execution.

Defense is in the Details: Every Node Accounted and Secured

On the sub-net, or across the Internet, *STTealth* Shield encrypts individual computer nodes utilizing multiple layers of rotating encryption algorithms, and changing keys at less than one-minute intervals. Encrypted journals report each node's activity to a master node. All inquiries and responses between nodes are also encrypted and recorded to an encrypted database. (No risky plaintext log files or parsing!) No PKI is required.

Multi-factor Time Varying Authentication

Each *STTealth* Shield system authenticates itself to other nodes, which in turn also authenticate themselves, using multiple unique factors and time-varying factors. Secure multi-factor authentication is the core of *STTealth* impenetrability. Errant nodes are black listed. This new dynamic authentication method is based on "temporary" factors, in addition to constant factors.

The *STTealth* Shield administrator can multiply authentication factors with a variety of new sources, including non-linear and environmental factors, and gaining the advantage to choose, change, and increase or decrease these factors, on demand. The administrator dictates the level of simplicity or complexity for the authentication process, as desired.

Node Communication and Management: Intuitive and Unassailable

STTealth Shield nodes communicate *only* with other *STTealth* Shield-authenticated nodes unless explicitly enabled. Strong distributed authentication and device subscription management identifies all nodes. Contaminated nodes are obstructed from further interaction with cleared nodes, shutting down malicious infiltration. Corrupt nodes are automatically blacklisted until re-authorized.

Through the *STTealth* Shield Management System (SSMS), the very small client application can be deployed to all computers within a network. The SSMS Administrator can then set up and manage relationships and privileges for every machine (users are unable to interact with SSMS). The SSMS is the UI that allows the SSMS Administrator to deploy and manage all computers, servers, and laptops in the network. The administrator may set up "zones" to represent access levels and can simply "drag and drop" the icon representing the individual computer into the zone to establish communication rights. The SSMS also provides set-up rules and rights for individual machines. An administrator can instantly cancel or modify rules, rights, and configurations; add new machines; and delete lost or stolen equipment from the subscription list, cancelling the ability of those machines to connect to the trusted network.

Weakness #5: *Managing Security is Difficult.* Most security approaches are reactive. *STTealth* Shield is pro-active. The complexity of protecting business environments with extensive geographies, business units, and functional areas with multiple objectives opens vulnerabilities in any security system.

STTealth Shield Management System: Counteracting Compromised Machines

In the event of lost or stolen equipment, the SSMS Administrator simply removes the item from all trusted subscriptions and it can no longer communicate through the *STTealth* Network. As new equipment is added or removed, the SSMS Administrator can easily update the network. All *STTealth* Shield enhanced entities can communicate with other *STTealth* Shield enhanced entities in the same seamless process as within their own *STTealth* Shield subscription networks. The SSMS Managers of each entity simply agree which machines can communicate across networks, and then they can set up the appropriate subscriptions. SSMS Managers can shut down their own network subscriptions to external communications at any time.

STTealth Shield Management System encrypts data-in-motion, data-at-rest, and individual computer nodes, securing all zones, unmanaged machines, and remote users with streamlined ease. The master node administrator is easy-to-use, highlighting a drag-and-drop interface. This agile and efficient security system operates with minimal overhead, low CPU utilization, and a small memory footprint.

Partner With STTealth
A Partner Program is available for *STTealth* Shield customers who need secure communications with non-*STTealth* Shield entities. The *STTealth* Shield Partner Program provides a limited *STTealth* Shield install on the non-*STTealth* Shield network for a small number of subscriptions communicating only between these two entities.

Subscription Security by Design
STTealth Shield incorporates a subscription methodology, which enforces a "contract" between two systems. The subscription consists of information about the system that it is contracted to communicate with (such as the address of the system) and which protocols the systems are allowed to use. Also included in the subscription is a randomly generated identity for the subscription, which is shared between the systems to facilitate in the validation of communications. The subscription identity changes automatically and is based on the number of times that the encryption keys for the subscription change. This can be as often as once per key change.

Many of the network security devices available today, include both virtual private network and firewall capabilities. *STTealth* Shield provides firewall capabilities in the form of three features: (1) Address Filters, (2) Service Filters, and (3) Protocol Filters. *STTealth* Shield eliminates relying on static encryption to secure communications, and employs a unique solution, which not only changes the keys used for encryption and decryption, but also changes the size of the key. The algorithm, used to generate the key, changes as often as every minute.

With a multi-point design, *STTealth* Shield successfully establishes a robust security environment that is not dependent on any single point of failure. In the event that an endpoint has to be taken offline, the remaining systems continue to communicate securely. A single management interface manages authorizations and subscriptions and simplifies the administration of security. Communications security is negotiated automatically, eliminating the need for managers to visit each computer system individually to update security protocols.

Authenticating with Unique Identifiers
The authentication system uses five or more unique identifiers. Several of these identifiers, which identify the individual device or machine, change on an irregular basis. All five (or more) parameters must match; otherwise, the traffic is ignored by *STTealth* Shield. As a result of the enhanced security concepts, the tools available to attackers are no longer effective, and prevent an attacker's ability to identify potential targets. Trojans are rendered useless, as outside systems cannot establish a communication channel with any Trojan introduced into the system. Using third party network analysis tools as a first alert system, if worms are released within a *STTealth* Shield environment, they are contained by restricting the communication subscriptions using the

administrative tools. Worms introduced from outside cannot get in, and worms introduced from inside cannot get out. *STTealth* Shield logs all communications traffic.

About STTealth Shield

A software company in Austin TX, STT LLC develops *STTealth* Shield[TM] an unparalleled security technique that secures all computer nodes and data on a network by concealing them from criminal detection. *STTealth* Shield safeguards computers from other computers within a sub-net, as well as masking partners from external Internet port probes. Subscribers effectively drop off the Radar[TM], providing no response to unauthenticated network messages inbound, and blocking all unauthorized outbound communication attempts.

STT LLC specializes in encrypted node and data security solutions protecting high-risk industries, such as DRM, film, banking/financial, legal, medical, military, retail/POS, and SCADA.

New EMP Protection Technology
Bill Joyce

Dr. Bill Joyce, *Chairman and CEO of Advanced Fusion Systems, and former CEO of Dow Chemical presented the most ambitious private sector electromagnetic pulse (EMP) testing and manufacturing facility in development. He leads the management and investment of $60M into this state-of-the-art facility that will produce vacuum tube technology that will assist electric grid operators to protect not only against ground-induced currents from solar storms or the E-3 pulses of HEMP, but also against the E-1 pulses of EMP whether produced by high-altitude nuclear burst covering large areas or pulsing devices that can disrupt or damage equipment at specific locations. The facility will be "capable of testing devices at line voltages up to 1.2 million VAC or VDC, under load conditions of up to 10 MW, and in a sub-100 picosecond rise time pulsed electric field environment of >250 kV/m. "One of the AFS product lines is the Bi-tron™, a bi-directional electron tube family designed for AC power electronics switching and control operations rated to 1.2 MV and current ratings in the hundreds of kiloamps. The Bi-tron is unique not only for the broader range of frequencies (of E-1 through E-3) they protect against (especially in the low frequency range), but also in the way it can reset and manage multiple events in rapid succession. The implications for protection as well as advanced weapons built on similar technology will undoubtedly compel the revision of current military specifications for at least some EMP protection technology.*

Dr. Joyce answered questions from the audience and provided a sober assessment of the need for technology solutions and his own commitment to provide some of them through the technology his firm is bringing to market.

BILL JOYCE: Good afternoon. Thank you for holding up through the day. I appreciate it. I'm Bill Joyce. I'm the CEO of Advanced Fusion Systems, AFS we call it. Hopefully this is— There we go. AFS offers EMP protective devices for transformers and generators and filtered feed throughs for shielded enclosures. We also manufacture a wide range of EMP sources with voltages up 250 kV per meter. I was quite pleased when John in his earlier talk mentioned some of our products, but I wasn't sure how I should react when he told me these sources are cheap and easy to make. We also offer some test services, including three large EMP cells that we're working on right at the moment.

A question when you're starting for protection, are you going to protect against E-1, E-3, and here are some things I think you need to think about. The fundamental physics governing the generation of a nuclear EMP pulse demands that if E-3 is present you're going to have E-1 and E-2 as well. The reverse isn't necessarily true. One can generate a non-nuclear EMP pulse that only has E-1, so caution is advised if you're designing the system. There is little sense in just protecting for E-3 in an environment where E-1 is anticipated coming down the path.

Numerous extremely powerful non-nuclear sources have been demonstrated. Some have been certified by the government with a field strength in excess of 250 kV per meter. Some of these devices are portable. This presents a threat which exceeds the levels of protection that are afforded by systems that just comply with MIL-188-125-A. These threats are relatively inexpensive, as John said, and lend themselves to multiple simultaneous attack scenarios.

The non-nuclear EMP threat arises from the ability to build extremely powerful radio transmitters that can duplicate the wave form and the intensities of EMP portion of a nuclear explosion, and the technology exists to build these transmitters of this nature that are portable and with many times the effects that a large nuclear EMP pulse will give. And shown in the picture here is a 35 kV per meter system that was built in collaboration program with the U.S. Army. As you see it fits easily into a small panel truck.

The best current available technology for E-1 protection consists of MOV devices and lightning gaps. Depending on these for protection is questionable in a nuclear EMP environment and unacceptable in a non-nuclear EMP environment. If the following graph sets the current technology and some AFS technology against the threats of these all in the same scale the black line is lightning, the dark red is an EMP pulse, the light red up above is a non-nuclear EMP pulse, and the blue line is where the protection starts with MIL-188. The green line is where the protection would start with a Bi-tron.

Clearly the MIL line looks like it takes all the steam out of lightning. A little questionable about the EMP pulse, but certainly not enough for the non-nuclear EMP, because remember that's a log scale so you're talking about things that a lot more powerful up above. The 188 is the current U.S. Government for high altitude EMP protection. It's widely used as design criteria, but the standard has some flaws in it. It utilizes a swept narrow band source for shield effectiveness testing, and that ignores the physics of the response of materials to an ultra-wide EMP pulse.

AFS will offer testing service that exceed MIL-188 and get a more realistic environment for both non-nuclear and nuclear EMP environments. The Bi-tron that was shown on the prior slide is a bidirectional electron tube designed for AC power electronic switching and control. It's a virtual family of tubes. It's suitable for all power electronic switching and control applications with voltages up to 1.2 million volts, and a current rating in the hundreds of kiloamps. For over voltage protection, and that includes EMP protection, in excess of MIL-188 the Bi-tron is designed to operate in the EMP environment and is capable of handling repeated pulses at a multi-kilohertz rep rate.

The size varies with the voltage that you're talking about. Units at 35 kV and down are 12 inches in diameter, 18 inches long. Units for 1,000 250 kV AC operation are approximately six foot in diameter. The systems provide an external contact signal to trip external protective devices. All the systems are self-resetting and capable of withstanding and protecting in an environment where there are repeated attacks in rapid succession. And of course by definition if they're able to do that they'll protect against lightning of all voltages.

This is a picture of one at 25 kV in two kA, 15 inches long and 12 inches in diameter. Obviously much bigger at 125 kV and two-and-a-half kA, about six foot in diameter and the length is equivalent to what you normally would have for an insulator at that kind of voltage.

AFS protection device for EMP is called EPS. Each of these is a dielectric vacuum enclosure, a ground conductor and some specialized internal structures. The devices implement the field collapse protocol. Detection and operation are autonomous. It protects first against E-1 and E-3. It provides hardened data output containing the information on the EPS status and EMP event alerts. The units are available from 41/60 to 1.2 MV.

This is a picture the same as John had shown earlier, an artist's concept of what they would be protecting, a large transformer. We have a 41/38 Bi-tron which has the same electrical characteristics, but it's built for, it's optimized for bulkhead mounting. It has a shielded protective feed through for the bulkhead. It exceeds the MIL specs. It's available in 275 kV and 250 kA and the tube is designed for EMP protection and transient suppression. And it looks like that.

AFS has some protective approaches for GMD. Obviously the field collapse EPS system that we talked about takes care of E-1 and E-3. We also have a neutral blocking device for a Kappenman method of GIC neutral blocking, and that's available as an integrated system through Phoenix Electric. It's also available as a switch itself to all qualified customers.

It looks like this at 35 kV and 200 kA, 24 inches long, 18 inches in diameter. I'm afraid we have a typo, because it's a lot heavier than that. This is our facility, manufacturing and office facility in Newtown, Connecticut. This is a satellite view. If you saw the building right now it has a lot more things on the roof and we're in construction on the one side, and one of the main elements being worked on now is an EMP and a GIC test facility.

It's essential that we test devices under realistic conditions, but we weren't able to find EMP test facilities capable of online testing devices up to a million volts. So we decided as part of our commitment to this arena we would construct some world-class EMP test facilities. The facility can inject simulated GIC signals to 25 kV DC and 100 kA. The facility will allow testing up to voltages of 1.2 million volts AC or DC, and under conditions that closely replicate what you would see in the real world, load conditions of 10 megawatts and sub-100 picosecond pulse electric feeds. So we think this moves us to something that will be quite practical and give us the kind of information that we need to know on these devices.

The cells we have *(are)* three cells. We're just finishing the steel lining on the first of them. It's 80 foot by 40 and 20 foot high. The second two cells are 135 by 50 and 50 feet high. They'll all test at 250 kV per meter. All cells can duplicate the magnetic field conditions of the largest transmission lines, digital instrumentation greater than 50 GS per second per channel and eight simultaneous channels and less than 20 picosecond resolution on the data.

This is some contact information. If I were you I would steer to Curtis Birnback, because he's the inventor of all this, but I'm happy to take your calls on anything that you might like to know.

CHUCK: Thank you very much. We should first start by giving him a quick round of applause, but I'd like to open it up for some questions.

CHUCK: I can imagine a few of you might have some, but I'd like to maybe basically start with one. Obviously this took a lot of commitment and investment to do this. It's interesting to see the private sector coming to the table to make this happen. Maybe you can give us a sense of the vision that you've had to have to make this come about and how this compares to your experience at the large Fortune 100 companies, and the difference between this and what this all might mean to the rest of us who are trying to observe that mix of activities that you've done over the years.

JOYCE: Well, a bunch of questions there, but to the first one, how long have we been at it and so long, Curtis actually worked on a lot of military applications over quite a few years, so these systems were a long time in developing, but they are proprietary products, i.e., presented to the military in a finished form. We put together the business as it stands right now. I started working on that about four years ago, and we'll have about 60 million invested. In this environment it's very, very difficult to get people to ante up, so we did it ourselves; less arguments that way.

But I think it's been a lot of fun. It's been very interesting, a lot of learning for me. I think that the other question is how different is it from being the CEO of a Fortune 500 company. Well, when you're buying something and you're the CEO of a Fortune 500 company you talk to somebody who talks to the realty department and they go out and search and find it. When you're doing what we did you get your shoes on and walk out the door and start looking yourself. So, you get an appreciation for a lot of jobs that have a lot more depth than you may ever have thought they did, and that's part of the fun of doing it.

CHUCK: Okay, do we have any other questions? I see hands, but I don't know; maybe people just scratching their heads. Mary has one. They'll bring the microphone to you.

MARY LASKY: How many people, how many utilities have actually invested with you?

JOYCE: Pardon?

MARY LASKY: How many utilities are actually anteing up for this?

CHUCK: As investors, I guess she was saying.

JOYCE: If you're talking about investors we did the investment ourselves. The response from the utilities has been very positive. In addition to EMP we have some current limiting devices and some other things that utilities, I would guess they probably have that on their front burner before EMP protection, and I think the issue is the one that one of the speakers talked about earlier. It's hard to get people interested in an event that has a terrible outcome, but it has a very low probability that it's going to happen in the near future. I think when people understand how, and I hate to say easily, but I'm going to use your words, how easily you can make one of these pulse devices I'm afraid we're going to find that that's not an event with low probability; it's an event with very high probability and it's not too far down the path.

CHUCK: I know we often talk about issues that are overwhelming and lots of people think of climate change, they think of peak oil, and at any particular point in time as they think about it when they're in a really tough economy they say, "Well this year is really bad, and maybe I have 50 years and my grandkids can figure it out or maybe I have 10 years." But what's interesting about these scenarios that we're looking at, it could be in five days or five months or 50 days or five years, not necessarily 50 years. And it's sort of like the sense of Hurricane Sandy, the storm is coming and it doesn't matter that times are hard and this is a tough problem; the storm is coming and so we need to do something about it.

JOYCE: I think the people in the Middle East will have this protection long before the United States does, because when you talk about climate change and you're in some place that the temperature is 130 degrees and all of a sudden your air conditioning is gone, man that's climate change.

CHUCK: Yes. Thank you very much. Oh we have a question.

MALE: Yes, I was going to ask you about commercial manufacturers of intentional EMP devices. I think it's now been publicly disclosed there is at least one. It's Boeing with their Champs drone. Do you know of other commercial manufacturers?

JOYCE: Well, we call ourselves commercial and Curtis manufactured these quite a few years ago, so they've been manufactured for a long period of time.

CHUCK: I see a hand, and this might be the last question unless another one pops up really fast, because we have a few other items to take care before we go.

DAVE GREER: I just have a quick question. Dave Greer, Air National Guard. You mentioned about the cost. Other people in the room all day were saying cost for protection is pretty low and you were joking about it. I was just wondering how much something, the protection would cost. Is it very expensive to implement something like this? I'm assuming yes. And is the Department of Defense interested in purchasing these things?

JOYCE: They aren't cheap. It's very dependent on the size, but if you're talking about a big transformer and you're protecting it it's a significant dollar amount, but if you look at it as a percentage of the transformer cost it's a very small amount. Obviously when we get down in size then they're much cheaper.

CHUCK: Thank you very much. Let's give him another round of applause.

Appendix

The following items have been included in the appendix to the InfraGard EMP SIG sessions of the Dupont Summit because of their direct bearing on technology policy presentations that appear in the Conference Proceedings. The appendix items will first be published on the Policy Studies Organization website at: http://www.ipsonet.org/conferences/the-dupont-summit/dupont-summit-2012 and may be subsequently placed on the secure InfraGard EMP SIG website.

They include the official notification of the proposed rule making on geomagnetic storms by FERC, a list of comments as of January 16, 2013, and a series of representative comments officially received on the FERC website for their Docket # 12-22, and subsequent Order No. 779 on "Reliability Standards for Geomagnetic Disturbances" (Issued May 16, 2013). Additional comments may be posted on the FERC site e-Library (ferc.gov) since then. It also includes material from the Federal Register concerning the companion ruling being undertaken by the Nuclear Regulatory Commission.

There is a brief section itemizing sample blogs commenting on cyber threats to information technology and industrial controls along with citizen concerns for emergency planning.

After that section, there are articles reflecting approaches to mitigating these infrastructure threats by conference presenters or their colleagues.

The final section includes the "after-action reports" from the first set of national level contingency planning workshops and exercises for impacts lasting a month or longer that have described by presenters at the conference. An economic impact assessment of a similar regional event is included along with an annotated bibliography and three pieces of proposed national legislation.

Appendix

FERC Notice of Proposed Rulemaking on GMD Official Documents and Comments

>   FERC Notice of Proposed Rulemaking and Related Press Releases

>   List of Comments Submitted to FERC as of January 16, 2013

>   143 FERC 61,147 FEDERAL ENERGY REGULATORY COMMISSION

>>   18 CFR Part 40 [Docket No. RM12-22-000; Order No. 779] Reliability Standards for Geomagnetic Disturbances (Issued May 16, 2013)

>   Overview of Nuclear Regulatory Commission Phased Ruling on GMD Impact to Nuclear Power Plant Safety

Key Comments Submitted to FERC

        Cynthia Ayers
        Chris Beck for EIS
        Curtis Birnbach for Adfusion
        Colorado Law Clinic
        Edison Electric Institute (THE AMERICAN PUBLIC POWER ASSOCIATION, THE EDISON ELECTRIC INSTITUTE, THE LARGE PUBLIC POWER COUNCIL AND THE NATIONAL RURAL ELECTRIC COOPERATIVE ASSOCIATION)
        William Kaewurt,
        John Kappenman for Storm Analysis Consultants
        Charles Manto on behalf of InfraGard National's EMP SIG
        North-American Energy Reliability Corporation
        Thomas Popik on behalf of Foundation for Resilient Societies, 2 Comments

Website and Blog Writings on High-impact Threats

        Mr. Larry Karisny (Discusses Sobel and Massey Cyber Initiatives)
        OurEnergyPolicy.org blog on FERC NOPR and Nuclear Regulatory Com'n
        Lanakila Washington on Proposed Emergency Measures
        Joseph Weiss on Declassified Report on Cyber Risks to Grid

Examples of Emerging Mitigating Technology

        Dr. Chris Beck, Testimony before Congress"
        Dr. George Baker, "National Infrastructure Protection Priorities for NEMP"
        Jackson, Noe, Baker "EM Field Threat Detector"
        Sobel, "ASBE Defeats Statistical Analysis and Cryptonalysis"
        Dr. Soysal, Soysal and Manto, EMP Protected Micro-grids"
        Dr. Clay Wilson, "Industrial and SCADA Systems Targeted for Cyber-attack"

GMD Contingency Planning Workshops and Exercises

        US Army War College Workshop on EMP "In the Dark", September 28-30, 2010
        National Defense University National GMD Exercise Oct 3-6, 2011
        (For the Secure Grid 11 Oct 4-5, 2011 report, see: http://www.ndu.edu/inss/docUploaded/Secure%20Grid%20'11%20After-Action%20Report.pdf
        General Kenneth Chrosniak debriefing of the NDU events of 2011

Economic Impact Assessment of Regional EMP Event, Sage Policy Group (reprint)

Bibliography for EMP and mitigation techniques, from EMP SIG web site

Proposed House Bills and Resolutions

>H Res 762
>Shield Act
>Grid Act
>Maine 126th Legislature 2013 H.P. 106, "An Act to Secure the Safety of Electrical
>Power Transmission Lines"

Conference Program

ADS AND ANNOUNCEMENTS

The following section contains ads and announcements. For additional information see the PSO website for updated exhibits and announcements: http://www.ipsonet.org/conferences/the-dupont-summit/dupont-summit-2012

The next Dupont Summit will be held on Friday December 6, 2013 at the Whittemore House in Washington, DC. For details on the InfraGard EMP SIG sessions of the 2013 Dupont Summit, contact cmanto@stop-EMP.com.

This section includes the draft of the newly proposed H.R. 2417 (the Shield Act) of the 113th Congress. It is similar to the Grid Act but with additional economic support for hardware-based solutions that the industry may use. New versions of these bills will be placed in the on-line exhibit section as they become available.

Emergency Law in Maine effective June 10, 2013

Resolve, Directing the Public Utilities Commission To Examine Measures To Mitigate the Effects of Geomagnetic Disturbances and Electromagnetic Pulse on the State's Transmission System

Emergency preamble. Whereas, acts and resolves of the Legislature do not become effective until 90 days after adjournment unless enacted as emergencies; and

Whereas, the North American Electric Reliability Corporation has identified 2013 as a peak year of solar activity that could result in a geomagnetic disturbance; and

Whereas, the impact of a significant geomagnetic disturbance or electromagnetic pulse on the reliability of Maine's electric grid is unknown; and

Whereas, the Public Utilities Commission may be able to identify measures to protect Maine's electric grid through a focused examination; and

Whereas, in the judgment of the Legislature, these facts create an emergency within the meaning of the Constitution of Maine and require the following legislation as immediately necessary for the preservation of the public peace, health and safety; now, therefore, be it

Sec. 1 Examination of vulnerabilities and mitigation. Resolved: That the Public Utilities Commission shall examine the vulnerabilities of the State's transmission infrastructure to the potential negative impacts of a geomagnetic disturbance or electromagnetic pulse capable of disabling, disrupting or destroying a transmission and distribution system and identify potential mitigation measures. In its examination, the commission shall:

1. Identify the most vulnerable components of the State's transmission system;

2. Identify potential mitigation measures to decrease the negative impacts of a geomagnetic disturbance or electromagnetic pulse;

3. Estimate the costs of potential mitigation measures and develop options for low-cost, mid-cost and high-cost measures;

4. Examine the positive and negative effects of adopting a policy to incorporate mitigation measures into the future construction of transmission lines and the positive and negative effects of retrofitting existing transmission lines;

5. Examine any potential effects of the State adopting a policy under subsection 4 on the regional transmission system;

6. Develop a time frame for the adoption of mitigation measures; and

7. Develop recommendations regarding the allocation of costs to mitigate the effects of geomagnetic disturbances or electromagnetic pulse on the State's transmission system and identify which costs, if any, should be the responsibility of shareholders or ratepayers; and be it further

Sec. 2 Monitor federal efforts regarding mitigation measures. Resolved: That the Public Utilities Commission shall actively monitor the efforts by the Federal Energy Regulatory Commission, the North American Electric Reliability Corporation, ISO New England and other regional and federal organizations to develop reliability standards related to geomagnetic disturbances and electromagnetic pulse; and be it further

Sec. 3 Report. Resolved: That the Public Utilities Commission shall report the results of its examination required pursuant to section 1 and the progress of regional and national efforts to develop reliability standards under section 2 to the Joint Standing Committee on Energy, Utilities and Technology by January 20, 2014. The Joint Standing Committee on Energy, Utilities and Technology may submit a bill to the Second Regular Session of the 126th Legislature based on the report.

Emergency clause. In view of the emergency cited in the preamble, this legislation takes effect when approved.

Foundation for Resilient Societies

Press Release: June 11, 2013

NEW LAW REQUIRES MAINE PUBLIC UTILITIES COMMISSION TO EXAMINE ELECTRIC GRID THREATS FROM SOLAR STORMS AND ELECTROMAGNETIC PULSE

MAINE LEADS THE NATION IN PROTECTING ITS ELECTRIC GRID

AUGUSTA ME—The State of Maine is the first in the nation to mandate a study of electric grid threats from solar storms and man-made electromagnetic pulse (EMP). A bill directing the Maine Public Utilities Commission (PUC) to examine measures to mitigate the effects of geomagnetic disturbances caused by solar storms and nuclear electromagnetic pulse on the state's electric grid transmission system was recently passed by a unanimous vote of the Maine House of Representatives and a 32-3 vote in the Senate. This new law requires the PUC to identify the most vulnerable components of the State's transmission system, and also identify potential protective measures, their costs, and probable timeframes to implement.

When sunspots erupt, they send masses of charged particles toward the earth. These particles interact with the earth's magnetic field causing so-called "geomagnetic disturbances". Solar storms can induce harmful currents in high voltage transmission lines. In March 1989, a relatively small solar storm plunged the Province of Quebec into a widespread grid blackout affecting six million electric customers. According to multiple U.S. Government reports, a five to ten-times larger solar storm—a historical example being the Carrington Event of 1859—could cause a years-long blackout for the State of

Current operational procedures of ISO-New England for solar storms are inadequate to protect the Maine and larger New England electric grids, according to testimony heard at a March 21 work session of the Energy, Utilities, and Technology Committee of the Maine State Legislature.

A nuclear bomb detonated high in the atmosphere could send an electromagnetic pulse downward and permanently damage electric grid transformers, according to reports from the Congressional Electromagnetic Pulse (EMP) Commission and Oak Ridge National Laboratories.

"I am proud that Maine has taken the lead on an issue that paralysis in Congress has let go unattended for too long," said Representative Andrea Boland of Sanford, sponsor of the bill. "Maine is particularly vulnerable to solar storms, due to its northerly latitude, geology, and proximity to the ocean. Legislators, the PUC, and the Governor understood what the experts were telling them, and responded."

The $1.4 billion Maine Power Reliability Program (MPRP), a high voltage transmission system upgrade, would substantially increase the risk of long-term electric grid blackout from solar storms, according to John Kappenman, one of the nation's foremost experts on the effects of sunspot eruptions on power grids. "While the new transmission lines add capacity, most of them parallel existing lines and terminate near existing substations," said Mr. Kappenman. "As a result, the redundant lines could result in a near-doubling of harmful currents induced by these storms in transmission lines in Maine. The antenna will be twice as big, and the storm impact to the power grid is likely to be twice as severe."

"Maine has narrowly missed blackouts due to solar storms, with numerous 'trips' of critical grid equipment," said Thomas Popik, Chairman of the Foundation for Resilient Societies, a non-profit group that conducts scientific studies on solar storms and their effect on electric grids. "Our recent study of the New England grid shows that generation and transmission capacity likely to be lost during a solar storm is several times the available reserves. We look forward to the upcoming study of the Maine PUC on this important issue."

For more background, the following information and contacts are provided:
Representative Andrea Boland of Sanford, Maine is available for interviews on the newly passed bill. Email Ms. Boland at sixwings@metrocast.net*, phone 207-432-7893.*

John Kappenman of Storm Analysis Consultants is available for interviews on technical aspects of solar storm effects on electric grids. Email Mr. Kappenman at jkappenma@aol.com*, phone 218-391-4015.*

The research report of the Foundation for Resilient Societies, "Solar Storm Risks for Maine and the New England Electric Grid, and Potential Protective Measures," *can be found on the foundation's website,* www.resilientsocieties.org*. Mr. Popik of the Foundation for Resilient Societies is available for interviews on costs of solar storm protection. Email Mr. Popik at* thomasp@resilientsocieties.org *, phone 603-321-1090.The Foundation for Resilient Societies does not endorse any specific legislative action in the State of Maine, or in other states, or in the U.S. Congress.*

FERC
FEDERAL ENERGY REGULATORY COMMISSION

News Release: May 16, 2013
Docket No. RM12-22-000
Item No. E-5

FERC Orders Development of Reliability Standards for Geomagnetic Disturbances

The Federal Energy Regulatory Commission (FERC) today issued a final rule requiring development of reliability standards that address the impact of geomagnetic disturbances (GMD) to ensure continued reliable operation of the nation's Bulk-Power System.

GMDs caused by solar events distort, with varying intensities, the earth's magnetic field. These events can have potentially severe, widespread effects on reliable grid operation, including blackouts and damage to critical or vulnerable equipment. Existing mandatory reliability standards do not adequately address GMD vulnerabilities on the Bulk-Power System.

Today's rule directs the North American Electric Reliability Corporation (NERC), the FERC-approved Electric Reliability Organization, to develop and submit new GMD standards in a two-stage process. The Commission did not require NERC to include any specific requirements in the GMD reliability standards; it identified issues to be considered and addressed in the standards development process.

In the first stage, NERC must file, within six months of the rule taking effect, one or more reliability standards requiring owners and operators of the Bulk-Power System to develop and implement operational procedures to mitigate GMD effects. The rule encourages implementation of the standards within six months of Commission approval.

The final rule also directs NERC to conduct a geomagnetic disturbance vulnerability assessment and identify facilities most at-risk from a severe disturbance.

In the second stage, NERC has 18 months to file standards identifying "benchmark GMD events," which define the severity of GMD events a

System. Those standards must require owners and operators to conduct initial and continuing assessments of the potential effects of specified "benchmark GMD events" on equipment and the Bulk-Power System as a whole. If the assessments identify potential effects from such events, the reliability standards should require a responsible entity to develop and implement plans to protect against instability, uncontrolled separation or cascading failures of the system.   The final rule takes effect 60 days after publication in the *Federal Register.*

R-13-20

143 FERC ¶ 61,147

UNITED STATES OF AMERICA

FEDERAL ENERGY REGULATORY COMMISSION

18 CFR Part 40

[Docket No. RM12-22-000; Order No. 779]

Reliability Standards for Geomagnetic Disturbances

(Issued May 16, 2013)

AGENCY: Federal Energy Regulatory Commission.

ACTION: Final Rule.

SUMMARY: Under section 215 of the Federal Power Act, the Federal Energy Regulatory Commission (Commission) directs the North American Electric Reliability Corporation (NERC), the Commission-certified Electric Reliability Organization, to submit to the Commission for approval proposed Reliability Standards that address the impact of geomagnetic disturbances (GMD) on the reliable operation of the Bulk-Power System. The Commission directs NERC to implement the directive in two stages.

In the first stage, NERC must submit, within six months of the effective date of this Final Rule, one or more Reliability Standards that require owners and operators of the Bulk-Power System to develop and implement operational procedures to mitigate the effects of GMDs consistent with the reliable operation of the Bulk-Power System. In the second stage, NERC must submit, within 18 months of the effective date of this Final Rule, one or more Reliability Standards that require owners and operators of the Bulk-Power System to conduct initial and on-going assessments of the potential impact of benchmark events on Bulk-Power System equipment and the Bulk-Power System as a whole.

The Second Stage GMD Reliability Standards must identify benchmark GMD events that specify what severity GMD events a responsible entity must assess for potential impacts on the Bulk-Power System. If the assessments identify potential impacts from benchmark GMD events, the Reliability Standards should require owners and operators to develop and implement a plan to protect against instability, uncontrolled separation, or cascading failures of the Bulk-Power System, caused by damage to critical or vulnerable Bulk-Power System equipment, or otherwise, as a result of a benchmark GMD event.

The development of this plan cannot be limited to considering operational procedures or enhanced training alone, but will, subject to the potential impacts of the benchmark GMD events identified in the assessments, contain strategies for protecting against the potential impact of GMDs based on factors such as the age, condition, technical specifications, system configuration, or location of specific equipment. These strategies could, for example, include automatically blocking geomagnetically induced currents from entering the Bulk-Power System, instituting specification requirements for new equipment, inventory management, isolating certain equipment that is not cost effective to retrofit, or a combination thereof.

EFFECTIVE DATE: This rule will become effective **[INSERT DATE 60 days after publication in the FEDERAL REGISTER (published May 23, 2013)]**.

The Task Force on National and Homeland Security

The Task Force on National and Homeland Security is a non-profit 501(c)3 and an official Congressional Advisory Board with direct access to key Members of Congress responsible for international and homeland security. The Task Force is the only organization with official standing that combines the areas of responsibility of both the Department of Defense and the Department of Homeland Security.

Our core mission is to educate policymakers and the American people on existential threats to the United States from nuclear and other weapons of mass destruction, and to support policies and to proactively initiate programs to protect the United States and its allies.

Task Force books "Electric Armageddon" and "Apocalypse Unknown" are both available through CreateSpace.com and Amazon.com.

Readying For Cyber Security

Three new federal laws and an Executive Order were enacted since January 2012 to require tighter controls in cyberspace. How they impact industry remains to be determined. However, the impact will be felt far and wide and will set the stage for significant new infrastructure and management systems in this realm. In the interest of preparing officials for what might come, this article aims to catalog and describe how these laws alter the present cyber landscape.

(1) 2012 National Defense Authorization Act § 818;

(2) 2013 National Defense Authorization Act, Subtitle D—Cyberspace-Related Matters § 933 & 935;

(3) Executive-Order 13636 Improving Critical Infrastructure Cyber-security; and

(3) H.R. 933 – The Consolidated and Further Continuing Appropriations Act, 2013 § 516 & 535.

Introduction

Beginning in October 2013 industry will begin to feel the impact of new cyber security requirements in the form of new permit requirements, certificates of operation, and other regulatory controls.

It should be noted that the provisions relating to the Department of Defense are likely to migrate to civilian agencies and industry in the absence of government sanctioned - industrial standards that go far enough to assure security. Voluntary industry standards including those developed and published by the ISO, the Open Group and NIST may be insufficient to meet security objectives so some in industry will encounter strict new regulatory requirements.

When viewed in their entirety these new mandates constitute a revision in policy regarding the cyber domain. The efficiencies offered by the Internet and new applications help managers to optimize their businesses, and secure profits, yet it is these systems that hold the mechanisms assailants use to attack, damage, destroy, injure and kill. These vulnerabilities constitute a new threat horizon that regulators are now required to reduce, and where possible, to eliminate. These mandates impose short compliance horizons, and enjoy non-partisan support in Congress, making them immune from political manipulation.

Public Law 112-81
Section 818: *Detection and Avoidance of Counterfeit Electronic Parts* sets out requirements governing the supply-chain for electronic components and IT systems that are sold to the Department of Defense. This provision shifts to suppliers the cost of tainted and counterfeit electronic parts, and the cost of rework or corrective action that may be required to remedy damage/losses resulting from their inclusion in systems.

The shift of liability to suppliers is a dramatic departure from contract policies of the last fifty years, and shifts significant burdens to product suppliers and systems integrators to ensure that only secure product is delivered to the customer and that suppliers indemnify the U.S. Government for components and systems that lead to failures. The ability to do this using the current global electronics and IT manufacturing base is nearly impossible, so a shift in production from Asia to the United States will certainly follow. Those firms that connect with the U.S. Government will be subject to similar requirements in order to push the threat of cyber attack as far away from the U.S. Government as is possible.

By way of example, those critical infrastructure elements, (e.g. oil, gas, electricity, banking, services, etc.) that connect to the U.S. Government for delivery of product, service, or reporting, will be required to make significant changes to protect U.S. Government assets

> *Other anticipated impacts include:*
>
> *(1) The ability of Commercial-Off-The-Shelf (COTS) suppliers to capitalize the risk;*
>
> *(2) The ability of COTS suppliers to certify to the integrity of their supply-chains for electronic components and IT Systems;*
>
> *(3) Constriction of eligible "trusted" supply chain participants; and*
>
> *(4) Cost of "trusted" COTS electronic and IT components and systems.*

and personnel.

DOD issued its initial Instruction (DODI 5200.44) to DOD Divisions and Agencies on November 5, 2012. The Instruction was followed in April, 2013 with additional Instructions, to be followed soon by detailed regulation.

While the final rules have not been issued, they are expected to require that each item sold to DOD will be subject to inspection at the point of manufacture to demonstrate that it is free of taint or counterfeit content. Taint is defined as the deliberate insertion of a functionality that was not specified by the (U.S.) designer in the reticles used to etch a microchip, or mechanisms by which chips are packaged, assembled onto motherboards and inserted in boxes for shipment to customers. New solicitations issued by the Department for electronic component and IT system procurements contain requirements for secure hardware. Contract awards for new secure systems is expected in the spring of 2014.

This requirement cannot be enforced abroad as it violates national sovereignty proscriptions. China, in particular, has been forceful in objecting to attempts by foreign firms seeking to enforce direct inspection of manufacturing facilities in this space, even those facilities that are owned by western firms.

DODI-5200.44 is expected to constrict the availability of "trusted COTS" electronic components and IT systems. The cost for trusted COTS systems will be more than that for commercial items, estimated to be about a factor of two, but much lower than the ten-fold factor for systems that must meet national security standards. As will be seen below, when discussing E.O. 13636, there is a high likelihood that demand for new electronic component and IT hardware by government and industry will emerge at the same time limiting inventory and putting upward pressure on cost.

Public Law P.L. 112-239

P.L. 112-239 extends the previous years national defense authorization regarding hardware and liability to software design and monitoring to assure the security of these systems as well. The two provisions most likely to impact industry are described below.

Section 933: *Improvements In Assurance Of Computer Software Procured By The Department Of Defense* requires development and implementation of a software assurance policy for the entire lifecycle of covered systems. Software assurance is defined as *"the level of confidence that a software program functions as intended and is free of vulnerabilities, either intentionally or unintentionally designed or inserted as part of the software, throughout the life cycle."*

Implementing this section requires:

> *(1) Deployment of automated vulnerability analysis tools in computer software code over the entire lifecycle of a covered system (e.g. development, operational testing, operations and sustainment phases, and retirement);*
>
> *(2) Identification/prioritization of security vulnerabilities and determination of appropriate remediation strategies; and*
>
> *(3) Remediation strategies are incorporated in contracts;*

Section 935: *Collection and Analysis Of Network Flow Data* requires development and demonstration of collection, processing, and storage technologies for network flow data that:

(1) Scale to the volume used by Tier 1 Internet Service Providers to collect and analyze the flow data across their networks;

(2) Reduce the cost and complexity of capturing and analyzing high volumes of flow data;

(3) Detect and identify cyber security threats, networks of compromised computers, and command and control sites used for managing illicit cyber operations and receiving information from compromised computers;

(4) Track illicit cyber operations for attribution of the source; and

(5) Provide early warning and attack assessment of offensive cyber operations.

Information of the kind described in items 3-5 is being aggressively pursued by the federal government. How they will acquire this information without new statutory authority is not clear.

Executive Order 13636

E.O. 13636 establishes a mandate to develop a Framework within which Government and Industry can work together to enhance security and resilience of the Nation's critical infrastructure and to <u>maintain a cyber environment that encourages efficiency, innovation, and economic prosperity while promoting safety, security, business confidentiality, privacy, and civil liberties.</u>

Sections 9 and 10 of the E.O. require:

(1) The identification of critical infrastructure elements where a cyber security incidence could reasonably result in catastrophic regional effects on public health and safety, economic security, or national security.

(2) Agencies with responsibility for regulating the security of critical infrastructure, and Independent Regulatory Agencies with responsibility for regulating the security of critical infrastructure shall determine if current cyber security regulatory requirements are sufficient given current and projected risks.

(3) Agencies shall make recommendations regarding increasing such authority to establish requirements to sufficiently address current and projected cyber risks.

The Administration will use existing authority under various Acts (e.g. environmental, food safety, pharmaceutical safety, energy, pipeline safety, transportation, nuclear safety, etc.) to compel critical infrastructure companies to secure their facilities and operations against cyber threats. Where accessibility to electronic components and IT systems by assailants is determined, companies will be subject to new regulations. The goal of the E.O. is to eliminate unauthorized access to critical infrastructure components entirely. To do this Companies will have to acquire new electronic, IT hardware, software, and will also require imposition of heightened cyber-hygiene procedures such as multi-factor authentication, limited access rules, and reporting to federal officials.

HR 933

The Consolidated and Further Continuing Appropriations Act, 2013 contains prohibitions regarding federal acquisition of electronic components and IT hardware from China.

> *(1) Section 516 prohibits the Departments of Commerce and Justice, NASA, NSF and OSTP from procuring any electronic or IT equipment from China unless the Department obtains certification from the FBI that it is in the national interest of the United States to do so.*
>
> *(2) Section 535 prohibits NASA and OSTP participation in bi-lateral activities, including the development of policies, programs, order, and contracts, with China or any Chinese owned Company.*

The mood in Congress is decidedly grim regarding the losses sustained by American industry, government and finance over the last three years. U.S. Cyber Command reported to Congress on numerous occasions that in 2010, 2011, and again in 2012, the cumulative losses to the U.S. amount to almost $1 trillion in each of these years. That amounts to 8.4% of GDP, an amount Congress has characterized as being unsustainable. Thus, despite considerable efforts to forestall prohibitions of the type contained here, this pattern of response has been cast.

These procurement prohibitions are being continued, and may be expanded in the FY '14 appropriations Bills. They are likely to mirror requirements contained in recently issued DOD solicitations, and lead to policies that can migrate to industries regulated by federal Departments and Agencies.

Limitations

Many of the leading firms in the electronic component and IT system market will have difficulty bringing new "more secure" product to market. Their supply chains run through many of the countries identified by the Congress and Administration as threatening the United States. Suppliers will have great difficulty in moving away from their supply chains to acquire new product. Many of these suppliers have for some time known, or suspected, that the electronic components and IT systems they have manufactured abroad were infected. As such they may be liable to their customers for knowingly selling such systems in the past. Thus, they will be reluctant to provide new more secure systems to customers, if by doing so they expose themselves to such claims.

New electronic component and IT System suppliers are emerging to meet market demand. Those who have moved into this realm, Harris Corporation, Lineage Technologies, LCC, etc., specialize in certain products, and secure supply chain controls. Whether they can meet the new market demand will depend upon their capacity.

Summary

Significant new law has been enacted within the last eighteen months regarding cyber security. These changes will be implemented using permits, licenses, and operations and management controls overseen by federal agencies. These changes will require that electronic component and IT systems be altered to make them impervious to attack through the Internet, cellular and radio networks. Current providers of such equipment will have to modify their products to make them secure from penetration, manipulation and sabotage. New market entrants are poised in this space to provide equipment where traditional firms cannot. At least for a short time this will impose significant expense across a broad spectrum of industries and businesses.

VENDOR CAPABILITIES

DUNS Number: 078304020

CAGE Code: 6LLW3

NAICS: 334112, 334118, 334210, 334413, 334418, 334419, 423430, 423690, 541614

Address: 1455 Pennsylvania Avenue Suite 400, Washington D.C. 20004

Phone Number: 703-899-4540

Email: info@lineagetech.com

www.lineagetech.com

Lineage Technologies LLC provides trustworthy versions of commercial-off-the-shelf computing and networking equipment that are manufactured in the United States with critical components sourced from US-based suppliers. The US government and critical infrastructure providers depend on commercial IT hardware built in a highly vulnerable global supply chain, creating a nearly insurmountable cybersecurity vulnerability. To protect against the burned-in/on-silicon "manufacturing cyber threat", Lineage teams with leading commercial vendors to license their designs, re-engineer them for the US supplier base, and deliver a state-of-the-art and trustworthy product. Beyond just the physical item, we deliver a full pedigree for the components, facilities, personnel and companies involved in the design, sourcing, and assembly of the product. In contrast to the government-specific vendors who deliver secure but trailing edge technologies at very high cost, Lineage delivers leading edge technologies at high security for far less than government-specific products.

A service-disabled veteran-owned business, Lineage's core competencies are in establishing supply chain pedigrees for the electronics manufacturing supply chain. This capability gives our customers the ability to make informed decisions about supply chain risk in their cybersecurity strategy. Working with some of the leading firms in the US technology sector, Lineage is able to deliver world-class technical solutions built in a high security environment to protect our customers from the manufacturing cyber threats pervasive in the industry. Leveraging the capabilities of our partners and suppliers, Lineage is able to deliver industry-leading design, fabrication, inventory management and assembly capabilities. Our organization's experience in supply chain

risk management, establishing high tech manufacturing supply chains and building inventory-tracking systems grows out of developing and managing two firms in particular: Verical, a leading marketplace for technology firms to trade pedigreed excess electronic component inventory, and Decernis, the largest dedicated supply chain management and tracking system in the food, chemical and pharmaceuticals industries. Together with our team's experience manufacturing world-class commercial technology hardware, this gives Lineage a compelling advantage in providing highly trustworthy products manufactured in pedigreed supply chains.

Lineage serves the entire United States from its main corporate office in downtown Washington, D.C. and its software development and operations teams located in Boulder, CO and Petaluma, CA. A new company founded in late 2011, Lineage has an experienced leadership team well suited to its mission. CEO and founder John P Brown has over a decade in operations and supply chain technology leadership and was co-founder of Verical, a marketplace for pedigreed electronic components now owned by Arrow Electronics. He served as an artillery officer in the United States Marine Corps and worked on the founding information assurance team at the Department of Homeland Security. He holds an AB in American history from Harvard, a masters of public administration from the Harvard Kennedy School and an MBA from the Harvard Business School. The Lineage team has decades of experience at some of the largest firms in the industry in defense technology, original component design and fabrication, telecom and computing product manufacturing, strategic sourcing, and supply chain system design and data operations.

113TH CONGRESS
1ST SESSION **H. R. 2417**

To amend the Federal Power Act to protect the bulk-power system and electric infrastructure critical to the defense and well-being of the United States against natural and manmade electromagnetic pulse ("EMP") threats and vulnerabilities.

IN THE HOUSE OF REPRESENTATIVES

JUNE 18, 2013

Mr. FRANKS of Arizona (for himself, Mrs. HARTZLER, Mr. POSEY, Mr. LAMBORN, Mr. KING of Iowa, Mr. BROUN of Georgia, Mr. PITTS, Mr. PITTENGER, Mr. LAMALFA, Ms. CLARKE, Mr. HUNTER, Mr. STEWART, Mr. WILSON of South Carolina, Mr. JORDAN, Mr. PERRY, Mr. GOSAR, Mr. DUNCAN of South Carolina, Mr. ROYCE, Mr. FORTENBERRY, and Mr. KLINE) introduced the following bill; which was referred to the Committee on Energy and Commerce, and in addition to the Committee on the Budget, for a period to be subsequently determined by the Speaker, in each case for consideration of such provisions as fall within the jurisdiction of the committee concerned

---

A BILL

To amend the Federal Power Act to protect the bulk-power system and electric infrastructure critical to the defense and well-being of the United States against natural and manmade electromagnetic pulse ("EMP") threats and vulnerabilities.

1   *Be it enacted by the Senate and House of Representa-*

2   *tives of the United States of America in Congress assembled,*

1 **SECTION 1. SHORT TITLE.**

2   This Act may be cited as the "Secure High-voltage

3 Infrastructure for Electricity from Lethal Damage Act"

4 or the "SHIELD Act".

5 **SEC. 2. FINDINGS.**

6   The Congress makes the following findings:

7   (1) According to the Report of the Commission

8   to Assess the Threat to the United States from

9   Electromagnetic Pulse Attack (in this Act referred

10   to as the "EMP Commission Report"), the society

11   and economy of the United States are "critically de-

12   pendent upon the availability of electricity.".

13   (2) According to the EMP Commission Report,

14   "continued electrical supply is necessary for sus-

15   taining water supplies, production and distribution

16   of food, fuel, communications, and everything else

17   that is part of our economy".

18   (3) According to the EMP Commission Report,

19   "contemporary U.S. society is not structured, nor

20   does it have the means, to provide for the needs of

21   nearly 300 million Americans without electricity.".

22   (4) According to the EMP Commission Report,

23   due to the existing electrical system operating at or

24   near its physical capacity, "a relatively modest upset

25   to the system can cause functional collapse.".

3

(5) According to the EMP Commission Report, electromagnetic pulse (in this Act referred to as "EMP") is a threat to the overall electrical power system.

(6) According to the EMP Commission Report, EMP occurs both naturally, such as geomagnetic storms, and via manmade devices.

(7) According to the EMP Commission Report, while the electric infrastructure "has a degree of durability against . . . the failure of one or a small number of [electric] components," the current strategy for recovery leaves the United States ill-prepared to respond effectively to an EMP attack that would potentially result in damage to vast numbers of components nearly simultaneously over an unprecedented geographic scale.

(8) According to the EMP Commission Report, EMP "may couple ultimately unmanageable currents and voltages into an electrical system routinely operated with little margin and cause the collapse of large portions of the electrical system.".

(9) According to the EMP Commission Report, a collapse of large portions of the electrical system will result in significant periods of power-outage and

1   "restoration from collapse or loss of significant por-

2   tions of the system [will be] exceedingly difficult.".

3   (10) According to the EMP Commission Re-

4   port, "should the electrical power system be lost for

5   any substantial period of time . . . the consequences

6   are likely to be catastrophic to civilian society.".

7   (11) According to the EMP Commission Re-

8   port, "the Commission is deeply concerned that

9   [negative] impacts [on the electric infrastructure]

10   are certain in an EMP event unless practical steps

11   are taken to provide protection for critical elements

12   of the electric system.".

13 **SEC. 3. AMENDMENT TO THE FEDERAL POWER ACT.**

14   (a) CRITICAL ELECTRIC INFRASTRUCTURE SECU-

15 RITY.—Part II of the Federal Power Act (16 U.S.C. 824

16 et seq.) is amended by adding after section 215 the fol-

17 lowing new section:

18 **"SEC. 215A. CRITICAL ELECTRIC INFRASTRUCTURE SECU-**

19   **RITY.**

20   "(a) DEFINITIONS.—For purposes of this section:

21   "(1) BULK-POWER SYSTEM; ELECTRIC RELI-

22   ABILITY ORGANIZATION; REGIONAL ENTITY.—The

23   terms 'bulk-power system', 'Electric Reliability Or-

24   ganization', and 'regional entity' have the meanings

1 given such terms in paragraphs (1), (2), and (7) of

2 section 215(a), respectively.

3   "(2) DEFENSE CRITICAL ELECTRIC INFRA-

4 STRUCTURE.—The term 'defense critical electric in-

5 frastructure' means any infrastructure located in the

6 United States (including the territories) used for the

7 generation, transmission, or distribution of electric

8 energy that—

9   "(A) is not part of the bulk-power system;

10 and

11   "(B) serves a facility designated by the

12 President pursuant to subsection (d)(1), but is

13 not owned or operated by the owner or operator

14 of such facility.

15   "(3) DEFENSE CRITICAL ELECTRIC INFRA-

16 STRUCTURE VULNERABILITY.—The term 'defense

17 critical electric infrastructure vulnerability' means a

18 weakness in defense critical electric infrastructure

19 that, in the event of a malicious act using an electro-

20 magnetic pulse, would pose a substantial risk of dis-

21 ruption of those electrical or electronic devices or

22 communications networks, including hardware, soft-

23 ware, and data, that are essential to the reliability

24 of defense critical electric infrastructure.

6

"(4) ELECTROMAGNETIC PULSE.—The term 'electromagnetic pulse' means 1 or more pulses of electromagnetic energy generated or emitted by a device capable of disabling, disrupting, or destroying electronic equipment by means of such a pulse.

"(5) GEOMAGNETIC STORM.—The term 'geomagnetic storm' means a temporary disturbance of the Earth's magnetic field resulting from solar activity.

"(6) GRID SECURITY THREAT.—The term 'grid security threat' means a substantial likelihood of—

"(A) a malicious act using an electromagnetic pulse, or a geomagnetic storm event, that could disrupt the operation of those electrical or electronic devices or communications networks, including hardware, software, and data, that are essential to the reliability of the bulk-power system or of defense critical electric infrastructure; and

"(B) disruption of the operation of such devices or networks, with significant adverse effects on the reliability of the bulk-power system or of defense critical electric infrastructure, as a result of such act or event.

3

1   "(7) GRID SECURITY VULNERABILITY.—The

2 term 'grid security vulnerability' means a weakness

3 that, in the event of a malicious act using an electro-

4 magnetic pulse, would pose a substantial risk of dis-

5 ruption to the operation of those electrical or elec-

6 tronic devices or communications networks, includ-

7 ing hardware, software, and data, that are essential

8 to the reliability of the bulk-power system.

9   "(8) LARGE TRANSFORMER.—The term 'large

10 transformer' means an electric transformer that is

11 part of the bulk-power system.

12   "(9) PROTECTED INFORMATION.—The term

13 'protected information' means information, other

14 than classified national security information, des-

15 ignated as protected information by the Commission

16 under subsection (e)(2)—

17   "(A) that was developed or submitted in

18   connection with the implementation of this sec-

19   tion;

20   "(B) that specifically discusses grid secu-

21   rity threats, grid security vulnerabilities, de-

22   fense critical electric infrastructure vulner-

23   abilities, or plans, procedures, or measures to

24   address such threats or vulnerabilities; and

8

1   "(C) the unauthorized disclosure of which

2   could be used in a malicious manner to impair

3   the reliability of the bulk-power system or of

4   defense critical electric infrastructure.

5   "(10) SECRETARY.—The term 'Secretary'

6   means the Secretary of Energy.

7   "(11) SECURITY.—The definition of 'security'

8   in section 3(16) shall not apply to the provisions in

9   this section.

10   "(b) EMERGENCY RESPONSE MEASURES.—

11   "(1) AUTHORITY TO ADDRESS GRID SECURITY

12   THREATS.—Whenever the President issues and pro-

13   vides to the Commission (either directly or through

14   the Secretary) a written directive or determination

15   identifying an imminent grid security threat, the

16   Commission may, with or without notice, hearing, or

17   report, issue such orders for emergency measures as

18   are necessary in its judgment to protect the reli-

19   ability of the bulk-power system or of defense critical

20   electric infrastructure against such threat. As soon

21   as practicable, but not later than 180 days after the

22   date of enactment of this section, the Commission

23   shall, after notice and opportunity for comment, es-

24   tablish rules of procedure that ensure that such au-

25   thority can be exercised expeditiously.

1 "(2) NOTIFICATION OF CONGRESS.—Whenever

2 the President issues and provides to the Commission

3 (either directly or through the Secretary) a written

4 directive or determination under paragraph (1), the

5 President (or the Secretary, as the case may be)

6 shall promptly notify congressional committees of

7 relevant jurisdiction, including the Committee on

8 Energy and Commerce of the House of Representa-

9 tives and the Committee on Energy and Natural Re-

10 sources of the Senate, of the contents of, and jus-

11 tification for, such directive or determination.

12 "(3) CONSULTATION.—Before issuing an order

13 for emergency measures under paragraph (1), the

14 Commission shall, to the extent practicable in light

15 of the nature of the grid security threat and the ur-

16 gency of the need for such emergency measures, con-

17 sult with the Secretary, other appropriate Federal

18 agencies, appropriate governmental authorities in

19 Canada and Mexico, the Electric Reliability Organi-

20 zation, and entities described in paragraph (4).

21 "(4) APPLICATION.—An order for emergency

22 measures under this subsection may apply to—

23   "(A) a regional entity; or

"(B) any owner, user, or operator of the bulk-power system or of defense critical electric infrastructure within the United States.

"(5) DISCONTINUANCE.—The Commission shall issue an order discontinuing any emergency measures ordered under this subsection, effective not later than 30 days after the earliest of the following:

"(A) The date upon which the President issues and provides to the Commission (either directly or through the Secretary) a written directive or determination that the grid security threat identified under paragraph (1) no longer exists.

"(B) The date upon which the Commission issues a written determination that the emergency measures are no longer needed to address the grid security threat identified under paragraph (1), including by means of Commission approval of a reliability standard under section 215 that the Commission determines adequately addresses such threat.

"(C) The date that is 1 year after the issuance of an order under paragraph (1).

"(6) COST RECOVERY.—If the Commission determines that owners, operators, or users of the

1 bulk-power system or of defense critical electric in-
2 frastructure have incurred substantial costs to com-
3 ply with an order under this subsection or subsection
4 (e) and that such costs were prudently incurred and
5 cannot reasonably be recovered through regulated
6 rates or market prices for the electric energy or
7 services sold by such owners, operators, or users, the
8 Commission shall, after notice and an opportunity
9 for comment, establish a mechanism that permits
10 such owners, operators, or users to recover such
11 costs.

12 "(e) MEASURES TO ADDRESS GRID SECURITY
13 VULNERABILITIES.—

14   "(1) COMMISSION AUTHORITY.—

15   "(A) RELIABILITY STANDARDS.—If the
16   Commission, in consultation with appropriate
17   Federal agencies, identifies a grid security vul-
18   nerability that the Commission determines has
19   not adequately been addressed through a reli-
20   ability standard developed and approved under
21   section 215, the Commission shall, after notice
22   and opportunity for comment and after con-
23   sultation with the Secretary, other appropriate
24   Federal agencies, and appropriate governmental
25   authorities in Canada and Mexico, issue an

1 order directing the Electric Reliability Organi-
2 zation to submit to the Commission for ap-
3 proval under section 215, not later than 30
4 days after the issuance of such order, a reli-
5 ability standard requiring implementation, by
6 any owner, operator, or user of the bulk-power
7 system in the United States, of measures to
8 protect the bulk-power system against such vul-
9 nerability. Any such standard shall include a
10 protection plan, including automated hardware-
11 based solutions. The Commission shall approve
12 a reliability standard submitted pursuant to
13 this subparagraph, unless the Commission de-
14 termines that such reliability standard does not
15 adequately protect against such vulnerability or
16 otherwise does not satisfy the requirements of
17 section 215.
18   ''(B) MEASURES TO ADDRESS GRID SECU-
19 RITY VULNERABILITIES.—If the Commission,
20 after notice and opportunity for comment and
21 after consultation with the Secretary, other ap-
22 propriate Federal agencies, and appropriate
23 governmental authorities in Canada and Mex-
24 ico, determines that the reliability standard
25 submitted by the Electric Reliability Organiza-

1 tion to address a grid security vulnerability
2 identified under subparagraph (A) does not
3 adequately protect the bulk-power system
4 against such vulnerability, the Commission shall
5 promulgate a rule or issue an order requiring
6 implementation, by any owner, operator, or user
7 of the bulk-power system in the United States,
8 of measures to protect the bulk-power system
9 against such vulnerability. Any such rule or
10 order shall include a protection plan, including
11 automated hardware-based solutions. Before
12 promulgating a rule or issuing an order under
13 this subparagraph, the Commission shall, to the
14 extent practicable in light of the urgency of the
15 need for action to address the grid security vul-
16 nerability, request and consider recommenda-
17 tions from the Electric Reliability Organization
18 regarding such rule or order. The Commission
19 may establish an appropriate deadline for the
20 submission of such recommendations.

21 ''(2) RESCISSION.—The Commission shall ap-
22 prove a reliability standard developed under section
23 215 that addresses a grid security vulnerability that
24 is the subject of a rule or order under paragraph
25 (1)(B), unless the Commission determines that such

14

1 reliability standard does not adequately protect
2 against such vulnerability or otherwise does not sat-
3 isfy the requirements of section 215. Upon such ap-
4 proval, the Commission shall rescind the rule pro-
5 mulgated or order issued under paragraph (1)(B)
6 addressing such vulnerability, effective upon the ef-
7 fective date of the newly approved reliability stand-
8 ard.

9 "(3) GEOMAGNETIC STORMS AND ELECTRO-
10 MAGNETIC PULSE.—Not later than 6 months after
11 the date of enactment of this section, the Commis-
12 sion shall, after notice and an opportunity for com-
13 ment and after consultation with the Secretary and
14 other appropriate Federal agencies, issue an order
15 directing the Electric Reliability Organization to
16 submit to the Commission for approval under section
17 215, not later than 6 months after the issuance of
18 such order, reliability standards adequate to protect
19 the bulk-power system from any reasonably foresee-
20 able geomagnetic storm or electromagnetic pulse
21 event. The Commission's order shall specify the na-
22 ture and magnitude of the reasonably foreseeable
23 events against which such standards must protect.
24 Such standards shall appropriately balance the risks
25 to the bulk-power system associated with such

15

events, including any regional variation in such
risks, the costs of mitigating such risks, and the priorities and timing associated with implementation. If
the Commission determines that the reliability
standards submitted by the Electric Reliability Organization pursuant to this paragraph are inadequate, the Commission shall promulgate a rule or
issue an order adequate to protect the bulk-power
system from geomagnetic storms or electromagnetic
pulse as required under paragraph (1)(B).

"(4) LARGE TRANSFORMER AVAILABILITY.—
Not later than 1 year after the date of enactment
of this section, the Commission shall, after notice
and an opportunity for comment and after consultation with the Secretary and other appropriate Federal agencies, issue an order directing the Electric
Reliability Organization to submit to the Commission for approval under section 215, not later than
1 year after the issuance of such order, reliability
standards addressing availability of large transformers. Such standards shall require entities that
own or operate large transformers to ensure, individually or jointly, adequate availability of large transformers to promptly restore the reliable operation of
the bulk-power system in the event that any such

•HR 2417 IH

1 transformer is destroyed or disabled as a result of

2 a geomagnetic storm event or electromagnetic pulse

3 event. The Commission's order shall specify the na-

4 ture and magnitude of the reasonably foreseeable

5 events that shall provide the basis for such stand-

6 ards. Such standards shall—

7   "(A) provide entities subject to the stand-

8   ards with the option of meeting such standards

9   individually or jointly; and

10   "(B) appropriately balance the risks asso-

11   ciated with a reasonably foreseeable event, in-

12   cluding any regional variation in such risks, and

13   the costs of ensuring adequate availability of

14   spare transformers.

15 "(d) CRITICAL DEFENSE FACILITIES.—

16   "(1) DESIGNATION.—Not later than 180 days

17 after the date of enactment of this section, the

18 President shall designate, in a written directive or

19 determination provided to the Commission, facilities

20 located in the United States (including the terri-

21 tories) that are—

22   "(A) critical to the defense of the United

23   States; and

1   "(B) vulnerable to a disruption of the sup-

2   ply of electric energy provided to such facility

3   by an external provider.

4   The number of facilities designated by such directive

5   or determination shall not exceed 100. The Presi-

6   dent may periodically revise the list of designated fa-

7   cilities through a subsequent written directive or de-

8   termination provided to the Commission, provided

9   that the total number of designated facilities at any

10   time shall not exceed 100.

11   "(2) COMMISSION AUTHORITY.—If the Commis-

12   sion identifies a defense critical electric infrastruc-

13   ture vulnerability that the Commission, in consulta-

14   tion with owners and operators of any facility or fa-

15   cilities designated by the President pursuant to

16   paragraph (1), determines has not adequately been

17   addressed through measures undertaken by owners

18   or operators of defense critical electric infrastruc-

19   ture, the Commission shall, after notice and an op-

20   portunity for comment and after consultation with

21   the Secretary and other appropriate Federal agen-

22   cies, promulgate a rule or issue an order requiring

23   implementation, by any owner or operator of defense

24   critical electric infrastructure, of measures to protect

25   the defense critical electric infrastructure against

1   such vulnerability. The Commission shall exempt

2   from any such rule or order any specific defense

3   critical electric infrastructure that the Commission

4   determines already has been adequately protected

5   against the identified vulnerability. The Commission

6   shall make any such determination in consultation

7   with the owner or operator of the facility designated

8   by the President pursuant to paragraph (1) that re-

9   lies upon such defense critical electric infrastructure.

10   "(3) COST RECOVERY.—An owner or operator

11   of defense critical electric infrastructure shall be re-

12   quired to take measures under paragraph (2) only to

13   the extent that the owners or operators of a facility

14   or facilities designated by the President pursuant to

15   paragraph (1) that rely upon such infrastructure

16   agree to bear the full incremental costs of compli-

17   ance with a rule promulgated or order issued under

18   paragraph (2).

19   "(e) PROTECTION OF INFORMATION.—

20   "(1) PROHIBITION OF PUBLIC DISCLOSURE OF

21   PROTECTED INFORMATION.—Protected information

22   shall—

23   "(A) be exempt from disclosure under sec-

24   tion 552(b)(3) of title 5, United States Code;

25   and

19

1   "(B) not be made available pursuant to

2 any State, local, or tribal law requiring disclo-

3 sure of information or records.

4 "(2) INFORMATION SHARING.—

5 "(A) IN GENERAL.—Consistent with the

6 Controlled Unclassified Information framework

7 established by the President, the Commission

8 shall promulgate such regulations and issue

9 such orders as necessary to designate protected

10 information and to prohibit the unauthorized

11 disclosure of such protected information.

12 "(B) SHARING OF PROTECTED INFORMA-

13 TION.—The regulations promulgated and orders

14 issued pursuant to subparagraph (A) shall pro-

15 vide standards for and facilitate the appropriate

16 sharing of protected information with, between,

17 and by Federal, State, local, and tribal authori-

18 ties, the Electric Reliability Organization, re-

19 gional entities, and owners, operators, and

20 users of the bulk-power system in the United

21 States and of defense critical electric infrastruc-

22 ture. In promulgating such regulations and

23 issuing such orders, the Commission shall take

24 account of the role of State commissions in re-

25 viewing the prudence and cost of investments

•HR 2417 IH

1   within their respective jurisdictions. The Com-

2   mission shall consult with appropriate Canadian

3   and Mexican authorities to develop protocols for

4   the sharing of protected information with, be-

5   tween, and by appropriate Canadian and Mexi-

6   can authorities and owners, operators, and

7   users of the bulk-power system outside the

8   United States.

9   "(3) SUBMISSION OF INFORMATION TO CON-

10   GRESS.—Nothing in this section shall permit or au-

11   thorize the withholding of information from Con-

12   gress, any committee or subcommittee thereof, or

13   the Comptroller General.

14   "(4) DISCLOSURE OF NONPROTECTED INFOR-

15   MATION.—In implementing this section, the Com-

16   mission shall protect from disclosure only the min-

17   imum amount of information necessary to protect

18   the reliability of the bulk-power system and of de-

19   fense critical electric infrastructure. The Commission

20   shall segregate protected information within docu-

21   ments and electronic communications, wherever fea-

22   sible, to facilitate disclosure of information that is

23   not designated as protected information.

24   "(5) DURATION OF DESIGNATION.—Informa-

25   tion may not be designated as protected information

1   for longer than 5 years, unless specifically redesig-

2   nated by the Commission.

3   "(6) REMOVAL OF DESIGNATION.—The Com-

4   mission may remove the designation of protected in-

5   formation, in whole or in part, from a document or

6   electronic communication if the unauthorized disclo-

7   sure of such information could no longer be used to

8   impair the reliability of the bulk-power system or of

9   defense critical electric infrastructure.

10   "(7) JUDICIAL REVIEW OF DESIGNATIONS.—

11   Notwithstanding subsection (f) of this section or sec-

12   tion 313, a person or entity may seek judicial review

13   of a determination by the Commission concerning

14   the designation of protected information under this

15   subsection exclusively in the district court of the

16   United States in the district in which the complain-

17   ant resides, or has his principal place of business, or

18   in the District of Columbia. In such a case the court

19   shall determine the matter de novo, and may exam-

20   ine the contents of documents or electronic commu-

21   nications designated as protected information in

22   camera to determine whether such documents or any

23   part thereof were improperly designated as protected

24   information. The burden is on the Commission to

25   sustain its designation.

1   "(f) JUDICIAL REVIEW.—The Commission shall act

2 expeditiously to resolve all applications for rehearing of

3 orders issued pursuant to this section that are filed under

4 section 313(a). Any party seeking judicial review pursuant

5 to section 313 of an order issued under this section may

6 obtain such review only in the United States Court of Ap-

7 peals for the District of Columbia Circuit.

8   "(g) PROVISION OF ASSISTANCE TO INDUSTRY IN

9 MEETING GRID SECURITY PROTECTION NEEDS.—

10   "(1) EXPERTISE AND RESOURCES.—The Sec-

11   retary shall establish a program, in consultation with

12   other appropriate Federal agencies, to develop tech-

13   nical expertise in the protection of systems for the

14   generation, transmission, and distribution of electric

15   energy against geomagnetic storms or malicious acts

16   using electromagnetic pulse that would pose a sub-

17   stantial risk of disruption to the operation of those

18   electronic devices or communications networks, in-

19   cluding hardware, software, and data, that are es-

20   sential to the reliability of such systems. Such pro-

21   gram shall include the identification and develop-

22   ment of appropriate technical and electronic re-

23   sources, including hardware, software, and system

24   equipment.

1   "(2) SHARING EXPERTISE.—As appropriate,

2 the Secretary shall offer to share technical expertise

3 developed under the program under paragraph (1),

4 through consultation and assistance, with owners,

5 operators, or users of systems for the generation,

6 transmission, or distribution of electric energy lo-

7 cated in the United States and with State commis-

8 sions. In offering such support, the Secretary shall

9 assign higher priority to systems serving facilities

10 designated by the President pursuant to subsection

11 (d)(1) and other critical-infrastructure facilities,

12 which the Secretary shall identify in consultation

13 with the Commission and other appropriate Federal

14 agencies.

15   "(3) SECURITY CLEARANCES AND COMMUNICA-

16 TION.—The Secretary shall facilitate and, to the ex-

17 tent practicable, expedite the acquisition of adequate

18 security clearances by key personnel of any entity

19 subject to the requirements of this section to enable

20 optimum communication with Federal agencies re-

21 garding grid security threats, grid security

22 vulnerabilities, and defense critical electric infra-

23 structure vulnerabilities. The Secretary, the Com-

24 mission, and other appropriate Federal agencies

25 shall, to the extent practicable and consistent with

1   their obligations to protect classified and protected

2   information, share timely actionable information re-

3   garding grid security threats, grid security

4   vulnerabilities, and defense critical electric infra-

5   structure vulnerabilities with appropriate key per-

6   sonnel of owners, operators, and users of the bulk-

7   power system and of defense critical electric infra-

8   structure.''.

9   (b) CONFORMING AMENDMENTS.—

10   (1) JURISDICTION.—Section 201(b)(2) of the

11   Federal Power Act (16 U.S.C. 824(b)(2)) is amend-

12   ed by inserting "215A," after "215," each place it

13   appears.

14   (2) PUBLIC UTILITY.—Section 201(e) of the

15   Federal Power Act (16 U.S.C. 824(e)) is amended

16   by inserting "215A," after "215,".

17 **SEC. 4. BUDGETARY COMPLIANCE.**

18   The budgetary effects of this Act, for the purpose of

19 complying with the Statutory Pay-As-You-Go Act of 2010,

20 shall be determined by reference to the latest statement

21 titled "Budgetary Effects of PAYGO Legislation" for this

22 Act, submitted for printing in the Congressional Record

23 by the Chairman of the House Budget Committee, pro-

1 vided that such statement has been submitted prior to the

2 vote on passage.

○

www.ingramcontent.com/pod-product-compliance
Lightning Source LLC
Chambersburg PA
CBHW051214200326
41519CB00025B/7113